Lecture Notes in Mathematics

A collection of informal reports and seminars
Edited by A. Dold, Heidelberg and B. Eckmann, Zürich

Series: Forschungsinstitut für Mathematik, ETH, Zürich · Adviser: K. Chandrasekharan

30

A. Frölicher · W. Bucher
Université de Genève

Calculus in Vector Spaces without Norm

1966

Springer-Verlag · Berlin · Heidelberg · New York

Work supported in part by the Swiss National Science Foundation

Library of Congress Catalog Card Number 66-30345. Printed in Germany. Title No. 7350

CONTENTS

INTRODUCTION

As emphasized by J. Dieudonné, calculus primarily deals with the approximation (in a neighborhood of some point) of given mappings of vector spaces by linear mappings. The approximating linear map has to be a "good" approximation in some precise sense: it has to be "tangent" to the given map. A very useful notion of "tangent" can easily be introduced for maps between normed vector spaces; it leads to the notion of "Fréchet-differentiable" mappings and gives, in particular for Banach spaces, a very satisfactory theory (cf. Chap. VIII of [3]).

It is well known that in this classical theory the notions of differentiability and derivative remain unchanged if one replaces the given norms by equivalent ones, i.e. by norms inducing the same topologies. It is natural therefore to look for a theory which does not use the norms, but only the topologies of the considered vector spaces. In fact, throwing out something which is irrelevant usually leads to a clarification and simplification on one side, and allows a more general theory on the other side. In the case of calculus, such a generalization is indeed desirable in view of applications to certain function spaces which have a natural topology, but no natural norm.

In classical theory, the norm is essentially used at two places: (1) One defines what maps $r: E_1 \longrightarrow E_2$ are tangent to zero at the origin (we simply call them "remainders") by means of the Fréchet-condition: $\lim_{x \to 0} \left(\frac{1}{\|x\|} \cdot \|r(x)\|\right) = 0$; (2) One defines a norm on the vector space $L(E_1;E_2)$, consisting of the continuous linear maps from E_1 into E_2, by taking, for $\ell \in L(E_1;E_2)$: $\|\ell\| = \sup_{\|x\| \leq 1} \|\ell(x)\|$. In order to obtain a similar theory for a class of non-normed topological vector spaces, one has therefore to choose essentially two definitions: (1) What are the remainders from E_1 to E_2;

(2) What is the topology of $L(E_1;E_2)$. The second definition comes in as soon as one wants to consider second (or higher) derivatives, since the first derivative f' of a (differentiable) map $f: E_1 \longrightarrow E_2$ is a map $f': E_1 \longrightarrow L(E_1;E_2)$. But all attempts which have been made along this line gave theories with a very serious deficiency: the composite of twice differentiable mappings did not turn out to be twice differentiable in general; in other words: there was no higher order chain rule. In fact, a look at the classical proof shows that the second order chain rule is a consequence of the first order chain rule and of the differentiability of the composition map $c: L(E_1;E_2) \times L(E_2;E_3) \longrightarrow L(E_1;E_3)$. But for non-normable topological vector spaces E_i there seems to be no separated topology on the spaces $L(E_i;E_j)$ such that the composition becomes differentiable.(*) Nevertheless, a way out of this difficulty was found: independently A. Bastiani and H.H. Keller realized that though there is no satisfactory topology on the spaces $L(E_i;E_j)$, there exist pseudo-topologies which have the desired properties. The authors are very much indebted to H.H. Keller for having drawn their attention to the fact that pseudo-topologies seem really the proper thing to use at this place.

(*) This statement is not very precise, in particular since it depends on the adopted definition of "differentiable". If, however, one requires that "differentiable" shall imply "continuous" and that the natural isomorphism between $L(\mathbf{R};E_i)$ and E_i shall be a homeomorphism, then one knows that with topologies one cannot succeed; in fact, the continuity of the composition map $c: L(\mathbf{R};E_1) \times L(E_1;E_2) \longrightarrow L(\mathbf{R};E_2)$ then is equivalent with the continuity of the evaluation map $e: L(E_1;E_2) \times E_1 \longrightarrow E_2$, and for non-normable spaces E_1, E_2 there is no topology on $L(E_1;E_2)$ for which this evaluation map e is continuous (cf. [7]).

The above remarks show that it is not for the sake of greatest possible generality that we develop our theory right from the beginning for pseudo-topological vector spaces (topological ones are special cases of these), but simply in order to obtain a satisfactory theory for a class of vector spaces containing at least some non-normable topological ones. In order to prove certain theorems of calculus, some restrictions however will have to be made: a class of pseudo-topological vector spaces, called "admissible" ones, will be introduced. This class contains in particular all separated locally convex topological vector spaces.

Since our whole theory works consistently with filters, §1 starts with some well known facts concerning filters. For a reader who is familiar with filters, it will be sufficient to have a look at (1.5.2); we found that at some places in the literature the inequality stated there was erroneously used as an equality. §2 presents the basic facts concerning pseudo-topologies and in particular pseudo-topological vector spaces. The material of sections 2.5 to 2.9 will not be used for the beginning of calculus and thus can be read later, whenever referred to.

§5 deals with what might be called the "mean value theorem". However, there is no mean value in it; but it is fundamental in the sense that it is used in order to prove practically all of the deeper results of calculus. We thus call it "fundamental theorem of calculus". Intuitively, it gives an estimate of the difference between the endpoints of a motion of a point in a vector space by means of the velocity of that motion, the estimation being made by means of convex sets. In the case of normed spaces, the theorem yields the well known estimate by means of the norm (cf. (8.5.1) of [3]) provided one chooses as convex set the closed unit ball; but being able to take other convex sets, we get better information also in this classical case: we not only can conclude that the point does not get too far if the velocity is not too big, but also that the point does get far, if the velocity is big (in the sense of lying in a multiple of the convex set in question). For later applications, some consequences of the theorem are established at the end of §5; in particular, two versions of the theorem in the form of filter inequalities will turn out to be useful. Another consequence is Taylor's formula; it will be given in a forthcoming publication.

In §7 one finds the definition of the admissible spaces and furthermore a result without which the theory would not be satisfactory: the class of admissible vector spaces is closed under the constructions used in calculus, yielding new spaces out of given ones, such as $L(E_1;E_2)$ or $C_k(E_1;E_2)$ out of E_1 and E_2.

In §8 we show that the relations between partial and total differentiability of a mapping of a direct product are as in classical theory; in particular, partial differentiability plus continuity of the partial derivatives implies total differentiability. We remark that this theorem uses in a very essential way the choice of the structure of the spaces $L(E_i;E_j)$, since "continuous" refers to the pseudo-topologies of the spaces in question.

The main results of §9 state that the p-th derivative at a point can be identified with a multilinear map which is symmetric, and that the composite of p-times differentiable maps is also p-times differentiable (p-th order chain rule).

The notion of a C_k-mapping introduced in §10 coincides with the usual notion of a k-times continuously differentiable mapping in the case of finite-dimensional spaces, while in general it is slightly more restrictive. The vector space consisting of the C_k-mappings from E_1 into E_2 is denoted by $C_k(E_1;E_2)$ or $C_k^*(E_1;E_2)$, depending on which of two pseudo-topological structures we consider (we always use one symbol to denote the space and its structure). The important spaces are the spaces $C_k^*(E_1;E_2)$; but for technical reasons it is useful to define them by means of the spaces $C_k(E_1;E_2)$ as auxiliary spaces and a general operator " $*$ " which associates to any pseudo-topology of a vector space a second one, having in addition a certain important property, called equability. In special cases, the operator "$*$" becomes the identity; in particular, if the spaces E_i are finite dimensional, the pseudo-topology of $C_k(E_1;E_2) = C_k^*(E_1;E_2)$ is nothing else than the topology of uniform convergence on bounded sets of the functions and their derivatives up to the k-th order. The case $k = \infty$ is obtained by forming a projective limit.

In §11 the differentiability and the C_p-nature of the composition map of C_k-mappings are investigated, the main results being theorems (11.2.21) and (11.2.26); here, the result stating that the composition map is of class C_p is in fact stronger than just saying that it is p-times continuously differentiable.

§12 deals with differentiable families of differentiable maps, "differentiable" now always meaning "differentiable of class C_∞". Having our theory of differentiation and also a pseudo-topology on the vector space of differentiable maps from E_1 into E_2, one can consider two sorts of differentiable families of such maps: a) A differentiable family of maps (depending, for instance, on a real parameter) is a differentiable map of $\mathbb{R} \times E_1$ into E_2; b) A differentiable family of maps is a differentiable map of $\mathbb{R}$ into the function space $C_\infty^*(E_1;E_2)$. The main result of §12 not only says that these two notions are completely equivalent, but even that the structures put on the space of all differentiable families according to either one of the two points of view a) or b) are the same; in other words, there is a canonical linear homeomorphism between $C_\infty^*(\mathbb{R} \times E_1;E_2)$ and $C_\infty^*(\mathbb{R};C_\infty^*(E_1;E_2))$. Moreover, the "parameter space" $\mathbb{R}$ can be replaced by any admissible equable vector space E. If we consider this isomorphism in the special case $E = E_1 = E_2 = \mathbb{R}$, for instance, then the space on the left hand side is classically well known, while on the right hand side we have a new function space, consisting of functions with values in the infinite dimensional space $C_\infty(\mathbb{R};\mathbb{R})$. Repeating this argument one sees that at least for many spaces E_i, E_j the set and the structure of $C_\infty^*(E_i;E_j)$ are uniquely determined if one requires the following two conditions: (1) in case E_i and E_j are finite dimensional, $C_\infty^*(E_i;E_j)$ is the set of classical C_∞-functions, with the topology of uniform convergence on compact sets of the functions and their derivatives; (2) the linear homeomorphism (12.2.5) mentioned before shall hold.

Depending on the choice of the two main definitions one obtains different theories. Our approach is different from those of A. Bastiani, H.H. Keller and E. Binz ([1],[6],[2]). In order to develop our theory, we always postulated that the definitions agree with the classical ones in the case of normed spaces, a condition which is not satisfied by the theories of A. Bastiani or of E. Binz. The structure of $L(E_1;E_2)$ defined by H.H. Keller for the case of locally convex spaces E_1, E_2 by means of families of semi-norms (cf. [5]), seems to be the same as the structure of our $L^*(E_1;E_2)$. In [6], H.H. Keller introduces various notions of differentiability and compares them with definitions that have been suggested by still other authors (cf. the references given there), restricting himself in that paper to locally convex spaces.

At the time being it is difficult to recognize which one of the various theories will eventually turn out to be the most useful one. That mainly depends on what theorems one gets and on what applications one wants to make. An implicit function theorem has not been obtained so far; in fact it is known that its classical formulation simply fails to hold. H.H. Keller has also established and motivated a series of basic properties that should hold in a useful theory of calculus (cf. [7]); we believe that our theory satisfies these conditions.

Throughout this report, we restrict ourselves to certain vector spaces; manifolds modelled on such vector spaces shall be considered later.

Though our notion of differentiability is a local property, a non-local condition is imposed on the so-called C_k-mappings; but this condition becomes trivial in the case of finite dimensional spaces, and, at least, it is not so restrictive as to rule out the identity map, as it would be the case if one had to restrict oneself to maps with compact or bounded support.

The first-named author has presented a first version of calculus for topological vector spaces in a Seminar of Professors A. Dold and B. Eckmann at the Swiss Federal Institute of Technology (ETH), Zurich, in summer 1963; it was not yet satisfactory, since there was no higher order chain rule. A part of the present theory was outlined by the same author in a series of lectures at the Forschungsinstitut für Mathematik of the ETH during the 1964/65 winter term.

The present work has been partially supported by the Swiss National Science Foundation.

§ 1. ELEMENTARY PROPERTIES OF FILTERS.

Since the whole theory is based on the convergence of filters, we recall here the fundamental facts concerning filters and state an inequality (in 1.5) which will be used very frequently in the sequel.

1.1. Filters and filter-basis.

A filter on a space (i.e. set) M is a non-empty set $\mathfrak{X}$ of subsets of M such that

(1) $\emptyset \notin \mathfrak{X}$ (*);

(2) $\left.\begin{array}{l} X_1 \in \mathfrak{X} \\ X_2 \supset X_1 \end{array}\right\} \longrightarrow X_2 \in \mathfrak{X}$;

(3) $X_1, X_2 \in \mathfrak{X} \longrightarrow X_1 \cap X_2 \in \mathfrak{X}$.

A filter-basis on M is a non-empty set $\mathfrak{B}$ of subsets of M such that

(1) $\emptyset \notin \mathfrak{B}$;

(2) $B_1, B_2 \in \mathfrak{B} \longrightarrow$ There exists $B_3 \in \mathfrak{B}$ with $B_3 \subset B_1 \cap B_2$.

Each filter is a filter-basis. Conversely, each filter-basis $\mathfrak{B}$ determines a filter $\mathfrak{X} = [\mathfrak{B}]$ as follows : $\mathfrak{X}$ consists of all subsets of M which contain a set of $\mathfrak{B}$. $\mathfrak{X}$ is called the filter generated by the filter-basis $\mathfrak{B}$. In particular, if B is any non-empty subset of M, $\mathfrak{B} = \{B\}$ is obviously a filter-basis. The filter

(*) $\emptyset$ always denotes the empty set

generated by it consists of all subsets of M containing the fixed set B and is denoted simply by $[B]$. Analogously, if $a \in M$, $[a]$ denotes the filter formed by the subsets of M containing the point a.

1.2. Comparison of filters on the same space.

The set of all filters on a given space M is partially ordered by the set-theoretic inclusion:

(1.2.1) $$\mathfrak{X}_1 \leq \mathfrak{X}_2 \longleftrightarrow \mathfrak{X}_1 \supset \mathfrak{X}_2 \quad (*).$$

We thus have the notions of infimum and supremum of a family $\{\mathfrak{X}_i\}_{i\in I}$ of filters on M : $\inf_{i\in I} \mathfrak{X}_i$ and $\sup_{i\in I} \mathfrak{X}_i$. The second always exists; it is the filter consisting of all sets of the form

(1.2.2) $$\bigcup_{i\in I} X_i \ , \text{ where } X_i \in \mathfrak{X}_i .$$

The first does not always exist; it will not be used. According to usual notation one also writes : $\sup(\mathfrak{X}_1, \mathfrak{X}_2) = \mathfrak{X}_1 \vee \mathfrak{X}_2$.

1.3. Mappings into direct products.

If $f_i : M_i \to N_i$, $i \in I$ resp. $i = 1,2$, are mappings, we denote by $\bigtimes_{i\in I} M_i$ the direct product of the sets M_i, and by $\bigtimes_{i\in I} f_i : \bigtimes_{i\in I} M_i \to \bigtimes_{i\in I} N_i$ resp. $f_1 x f_2 : M_1 x M_2 \to N_1 x N_2$ the maps defined as follows:

(*) In [4] Fischer uses the symbol "$\leq$" in the other sense; definition (1.2.1) is the one used e.g. by Kowalsky in [b].

(1.3.1) $(\underset{i\in I}{\times} f_i)(\{x_i\}_{i\in I}) = \{f_i(x_i)\}_{i\in I}$ resp. $(f_1 \times f_2)(x_1,x_2) = (f_1(x_1),f_2(x_2))$.

In the special case where $M_i = M$ for all $i\in I$, $\underset{i\in I}{\times} M_i$ is usually denoted by M^I, and we further denote by

$$\prod_{i\in I} f_i : M \longrightarrow \underset{i\in I}{\times} N_i \quad \text{resp.} \quad [f_1,f_2] : M \longrightarrow N_1 \times N_2$$

the maps defined as follows:

(1.3.2) $$(\prod_{i\in I} f_i)(x) = \{f_i(x)\}_{i\in I} \quad \text{resp.} \quad [f_1,f_2]\,(x) = (f_1(x),f_2(x)).$$

These maps are related by means of the diagonal maps

$$d : M \longrightarrow M^I \quad \text{resp.} \quad d : M \longrightarrow M \times M = M^2$$

as follows :

(1.3.3) $$\prod_{i\in I} f_i = (\underset{i\in I}{\times} f_i)\circ d \quad \text{resp.} \quad [f_1,f_2] = (f_1 \times f_2)\circ d.$$

1.4. Images of filters under mappings.

Let $f: M \longrightarrow N$ be a mapping and $\mathfrak{X}$ a filter on M. The set $\{f(X) \mid X \in \mathfrak{X}\}$ is than a filter-basis on N, which generates a filter denoted by $f(\mathfrak{X})$ and called the image of $\mathfrak{X}$ under the mapping f. The use of the same symbol f is justified because the functor in question is covariant :

(1.4.1) If $M \xrightarrow{f} N \xrightarrow{g} P$, then $(g\circ f)(\mathfrak{X}) = g(f(\mathfrak{X}))$;

and also because

(1.4.2) $$f([x]) = [f(x)].$$

The induced mapping for filters is order-preserving :

(1.4.3) $\mathfrak{X}_1 \leq \mathfrak{X}_2 \implies f(\mathfrak{X}_1) \leq f(\mathfrak{X}_2)$.

Let now $\mathfrak{X}_i$ be a filter on M_i, $i = 1,2$. Then we define:

(1.4.4) $\mathfrak{X}_1 \times \mathfrak{X}_2$ is the filter on $M_1 \times M_2$ generated by the following filter-basis : $\{X_1 \times X_2 \mid X_1 \in \mathfrak{X}_1;\ X_2 \in \mathfrak{X}_2\}$.

If further $g : M_1 \times M_2 \longrightarrow N$ is a map, then:

(1.4.5) $g(\mathfrak{X}_1, \mathfrak{X}_2)$ denotes the filter generated by the following filter-basis: $\{g(X_1, X_2) \mid X_1 \in \mathfrak{X}_1; X_2 \in \mathfrak{X}_2\}$.

It follows easily that

(1.4.6) $g(\mathfrak{X}_1, \mathfrak{X}_2) = g(\mathfrak{X}_1 \times \mathfrak{X}_2)$.

1.5. An inequality between images of filters.

Let $\mathfrak{X}$ be a filter on M and $d : M \longrightarrow M \times M$ be the diagonal map : $d(x) = (x,x)$. Then

(1.5.1) $d(\mathfrak{X}) \leq \mathfrak{X} \times \mathfrak{X}$.

Proof. Let $A \in \mathfrak{X} \times \mathfrak{X}$. Then $A \supset X_1 \times X_2$, where $X_1, X_2 \in \mathfrak{X}$. With $X = X_1 \cap X_2$ we have: $A \supset X \times X$, where $X \in \mathfrak{X}$. But since $X \times X \supset d(X)$, it follows that $A \supset d(X)$ which shows that $A \in d(\mathfrak{X})$.

Usually, this inequality will be used in combination with mappings. A typical example is as follows.

(1.5.2) Let $h_i : M \longrightarrow N_i$, $i = 1,2$ and $g : N_1 \times N_2 \longrightarrow P$ be maps. If $f : M \longrightarrow P$ is the map defined by $f(x) = g(h_1(x), h_2(x))$, then $f(\mathfrak{X}) \leq g(h_1(\mathfrak{X}), h_2(\mathfrak{X}))$.

<u>Proof</u>. Since by (1.3.3) we have $f - g \circ [h_1, h_2] = g \circ (h_1 \times h_2) \circ d$, the inequality follows from (1.5.1), using (1.4.3), (1.4.6) and the equality $(h_1 \times h_2)\ (\mathfrak{X} \times \mathfrak{X}) = h_1(\mathfrak{X}) \times h_2(\mathfrak{X})$.

We shall refer to (1.5.2) whenever we have occasion to use an inequality of this type, even though the situation may be somewhat different. For example it could be that there are several variables, or the right hand side might contain x repeated more than once.

§ 2. PSEUDO-TOPOLOGICAL VECTOR SPACES.

For a detailed introduction of the notion of pseudo-topological spaces, the reader is referred to [4] . We shall introduce a slightly different notation which will be more convenient for our purpose. Some associated structures will be introduced, and questions concerning continuity will be discussed. It turned out that a condition which is slightly stronger than ordinary continuity will play an important role: the notion of equable continuity. In particular we investigate the case of linear and multilinear maps.

2.1. Pseudo-topological spaces.

A pseudo-topology (or limit-structure) on a space (i.e. set) $\mathfrak{M}$ consists in assigning to each $x \in \mathfrak{M}$ a set of filters on $\mathfrak{M}$, such filters being described as "converging to x". The following axioms are supposed satisfied: (1) If a filter converges to x, then so does any smaller filter; (2) If two filters converge to x, then so does their supremum; (3) The filter $[x]$ converges to x.

A pseudo-topological space E consists of a set $\mathfrak{M}$ together with a pseudo-topology on $\mathfrak{M}$. The set $\mathfrak{M}$ is called the underlying space and will be denoted by $\underline{E}$. If a filter $\mathfrak{X}$ on $\underline{E}$ converges to x in the sense of the given pseudo-topology, we write $\mathfrak{X} \downarrow_x E$ ($\mathfrak{X}$ converges to x on E). The axioms can now be expressed in the following way.

(2.1.1) For all $x \in \underline{E}$ one has :

$$(1) \quad \left.\begin{array}{l} \mathfrak{X} \downarrow_x E \\ \mathfrak{Y} \leq \mathfrak{X} \end{array}\right\} \Rightarrow \mathfrak{Y} \downarrow_x E ;$$

$$(2) \quad \mathfrak{X}_i \downarrow_x E \text{ for } i = 1,2 \longrightarrow \mathfrak{X}_1 \vee \mathfrak{X}_2 \downarrow_x E ;$$

$$(3) \quad [x] \downarrow_x E.$$

Topological spaces can be considered as special cases of pseudo-topological spaces. In fact, if E is a topological space, we define :

$$\mathfrak{X} \downarrow_x E \longleftrightarrow \mathfrak{X} \leq \mathcal{U}_x ,$$

where $\mathcal{U}_x$ is the filter of neighborhoods of x. Knowing the converging filters on E, conversely we can recover the topology, since then $\mathcal{U}_x = \sup_{\mathfrak{X} \downarrow_x E} \mathfrak{X}$. Thus, a necessary condition for a pseudo-topological space to be a topological one is the following: $(\sup_{\mathfrak{X} \downarrow_x E} \mathfrak{X}) \downarrow_x E$. This condition is not in general sufficient: however we shall see in (2.4) that it is sufficient in the case in which we shall be interested, i.e. if E is a pseudo-topological vector space. For details, see [4] .

2.2. Continuity.

Let E_1, E_2 be pseudo-topological spaces and $f: \underline{E}_1 \longrightarrow \underline{E}_2$ a map. We say that $f: E_1 \longrightarrow E_2$ is <u>continuous at the point</u> $a \in E_1$ iff (*)

$$\mathfrak{X} \downarrow_a E_1 \Longrightarrow f(\mathfrak{X}) \downarrow_{f(a)} E_2 .$$

$f: E_1 \longrightarrow E_2$ continuous means: continuous at each point. In the topological case the definition is equivalent to the usual one. One further verifies easily that the composite of continuous maps is also continuous.

$f: E_1 \longrightarrow E_2$ is a homeomorphism means that $f: \underline{E}_1 \longrightarrow \underline{E}_2$ is bijective and that $f: E_1 \longrightarrow E_2$ as well as $f^{-1}: E_2 \longrightarrow E_1$ are continuous.

2.3. Induced structures.

One can introduce a partial ordering on the set of pseudo-topologies of a fixed space. Let E_1, E_2 be two pseudo-topological spaces with $\underline{E}_1 = \underline{E}_2$.

(2.3.1) Definition: The structure of E_1 is called <u>finer</u> than that of E_2, and we write $E_1 \leq E_2$, iff

$$\mathfrak{X} \downarrow_x E_1 \Longrightarrow \mathfrak{X} \downarrow_x E_2 .$$

(*) From time to time we shall make use of the expression "iff"; as usual it stands for "if and only if".

We also say the structure of E_2 is <u>coarser</u> than that of E_1. This is equivalent to the condition: the identity map $i : E_1 \longrightarrow E_2$ is continuous.

As in the topological case, one can define structures which are induced by mappings. We will use only the following case: Let E_i, $i \in I$, be pseudo-topological spaces, $\mathfrak{M}$ a space (without structure), and

$$f_i : \mathfrak{M} \longrightarrow \underline{E}_i \quad , \; i \in I \; .$$

Then there exists on $\mathfrak{M}$ a unique pseudo-topology which is caracterized as being the coarsest pseudo-topology on $\mathfrak{M}$ such that all maps $f_i : \mathfrak{M} \longrightarrow E_i$ are continuous. Denoting $\mathfrak{M}$ together with that structure by E, we have ($\mathfrak{X}$ being a filter on $\mathfrak{M}$) :

(2.3.2) $$\mathfrak{X} \downarrow_x E \iff f_i(\mathfrak{X}) \downarrow_{f_i(x)} E_i \quad \text{for all } i \in I.$$

In fact, one easily verifies that with this definition the axioms (2.1.1) are satisfied; and (2.3.2) certainly defines the coarsest structure on $\mathfrak{M}$ such that all maps $f_i : \mathfrak{M} \longrightarrow E_i$ get continuous.

If all the spaces E_i have the property that for each point $x \in E_i$ one has $\sup\limits_{\mathfrak{X} \downarrow_x E_i} \mathfrak{X} \downarrow_x E_i$, then also E has this property. This is easily verified, using that for any map $f : \mathfrak{M} \longrightarrow N$ and any family $\{\mathfrak{X}_j\}_{j \in J}$ of filters on $\mathfrak{M}$, one has

(2.3.3) $$f(\sup_{j\in J} \mathfrak{X}_j) = \sup_{j\in J} f(\mathfrak{X}_j) ;$$

which is an immediate consequence of the set-theoretic equality

$$f(\bigcup_{j\in J} X_j) = \bigcup_{j\in J} f(X_j) .$$

We consider now three special cases of structures induced by mappings : subspaces, direct products and projective limits.

a) Each subset A of a pseudo-topological space E has a natural pseudo-topology, namely the one induced by the inclusion map $i : A \longrightarrow E$. A together with this structure is called a <u>subspace</u> of E. Denoting the subspace thus obtained by E_1, we have according to (2.3.2) :

(2.3.4) $$\mathfrak{X} \downarrow_x E_1 \iff i(\mathfrak{X}) \downarrow_x E .$$

b) Given a family $\{E_i\}_{i\in I}$ of pseudo-topological spaces, we call <u>direct product</u> and denote by $\times_{i\in I} E_i$ the pseudo-topological space whose underlying set is the direct product of the sets $\underline{E}_i$, together with the structure induced by the projections

$$\pi_k : \times_{i\in I} \underline{E}_i \longrightarrow E_k , \quad k\in I.$$

Thus (2.3.2) yields for this case:

(2.3.5) $$\mathfrak{X} \downarrow_x (\times_{i\in I} E_i) \iff \pi_k(\mathfrak{X}) \downarrow_{\pi_k(x)} E_k \text{ for all } k\in I.$$

In case of finite direct products we write $E_1 \times \ldots \times E_n$ instead of $\times_{i\in\{1,\ldots,n\}} E_i$.

c) Let a projective system of pseudo-topological spaces be given, i.e. to each element i of a directed set $\{I, \geqslant\}$ a pseudo-topological space E_i is associated, such that for $i_1 \geqslant i_2$, $\underline{E}_{i_1}$ is a subset of $\underline{E}_{i_2}$, with a continuous inclusion map. Then $E = \text{proj.lim}_{i \in I} E_i$ is the pseudo-topological space whose underlying set $\underline{E}$ is the intersection $\underline{E} = \bigcap_{i \in I} \underline{E}_i$ and whose pseudo-topology is the one induced by the inclusion maps $j_i : \underline{E} \longrightarrow E_i$.

In the case of subspaces and direct products, the following two lemmas will be used later.

(2.3.6) Lemma. Let $f : E_1 \longrightarrow E_2$ be a map and suppose that E_2 is a subspace of E_3, $i : E_2 \longrightarrow E_3$ being the inclusion map. Then $f : E_1 \longrightarrow E_2$ is continuous if and only if $i \circ f : E_1 \longrightarrow E_3$ is continuous.

Proof. Necessity is obvious, since $i : E_2 \longrightarrow E_3$ is continuous. Suppose conversely that $i \circ f$ is continuous, and let $\mathfrak{X} \downarrow_a E_1$. Then $(i \circ f)(\mathfrak{X}) \downarrow_{(i \circ f)(a)} E_3$, or equivalently (by (1.4.1)): $i(f(\mathfrak{X})) \downarrow_{f(a)} E_3$. By (2.3.4) this yields $f(\mathfrak{X}) \downarrow_{f(a)} E_2$, which proves the continuity of f.

(2.3.7) Lemma. Let $\mathfrak{X}$ be a filter on $E_1 \times E_2$. Then $\mathfrak{X} \downarrow_x E_1 \times E_2 \Longleftrightarrow$ there exist $\mathfrak{X}_1$, $\mathfrak{X}_2$ with $\mathfrak{X}_i \downarrow_{\pi_i(x)} E_i$, i = 1,2, and $\mathfrak{X} \leq \mathfrak{X}_1 \times \mathfrak{X}_2$.

Proof. 1) Since for any subset $X \subset E_1 \times E_2$ one has $X \subset \pi_1(X) \times \pi_2(X)$, one has for any filter $\mathfrak{X}$ on $E_1 \times E_2$:

(2.3.8) $$\mathfrak{X} \leq \pi_1(\mathfrak{X}) \times \pi_2(\mathfrak{X}).$$

Hence, if $\mathfrak{X} \downarrow_x E_1 \times E_2$, we can choose $\mathfrak{X}_i = \pi_i(\mathfrak{X})$.

2) If, conversely, $\mathfrak{X} \leq \mathfrak{X}_1 \times \mathfrak{X}_2$, where $\mathfrak{X}_i \downarrow_{\pi_i(x)} E_i$, then $\pi_i(\mathfrak{X}) \leq \pi_i(\mathfrak{X}_1 \times \mathfrak{X}_2) = \mathfrak{X}_i$, and according to (2.3.5) this proves that $\mathfrak{X} \downarrow_x E_1 \times E_2$.

2.4. Pseudo-topological vector spaces.

Let the underlying space $\underline{E}$ of a pseudo-topological space E be a vector space (*). The pseudo-topology is called compatible with the vector space structure (shortly: compatible) if the maps

$$E \times E \xrightarrow{+} E,$$

$$\mathbb{R} \times E \xrightarrow{\cdot} E$$

are continuous. By $\mathbb{R}$ we always denote the reals, taken with the usual topology.

A pseudo-topological vector space is a vector space together with a compatible pseudo-topology on it. More precisely: it is a triple, consisting of a set together with two structures on it, namely an algebraic structure (of vector space) and a pseudo-topological structure.

(*) Vector space here always means: vector space over the reals $\mathbb{R}$.

Continuity of addition implies that the translations are homeomorphisms. Therefore:

(2.4.1) $\mathfrak{X}\downarrow_a E \Longleftrightarrow \mathfrak{X}-a\downarrow_0 E$ or equivalently

$\mathcal{Y}+a\downarrow_a E \Longleftrightarrow \mathcal{Y}\downarrow_0 E.$

By $\mathfrak{X}$-a we denote of course the image of $\mathfrak{X}$ under the translation map $x \longmapsto x-a$. More generally, we write $g(\mathfrak{X},\mathcal{Y}) = \mathfrak{X}-\mathcal{Y}$ if $g(x,y) = x-y$. Then one easily verifies that $\mathfrak{X}$-a can also be considered as the image of two filters under the difference map: $\mathfrak{X}-a = \mathfrak{X}-[a]$. Similarly for $\mathcal{Y}+a$: $\mathcal{Y}+a = \mathcal{Y}+[a]$.

In view of (2.4.1) the pseudo-topology of a pseudo-topological vector space E is completely known if we know what filters converge to zero. Hence we only need one relation, that of "converging to zéro", and we shall simplify the notation by writing $\mathfrak{X}\downarrow E$ instead of $\mathfrak{X}\downarrow_0 E$. Thus "$\downarrow$..." simply means : "converges to zero in...".

The continuity of the operations implies the following <u>compatibility conditions</u>:

(2.4.2)
(1) $\mathfrak{X}_1\downarrow E,\ \mathfrak{X}_2\downarrow E \longrightarrow \mathfrak{X}_1+\mathfrak{X}_2\downarrow E;$

(2) $\mathfrak{X}\downarrow E,\ \lambda\in\mathbb{R} \longrightarrow \lambda\cdot\mathfrak{X}\downarrow E;$

(3) $\mathfrak{X}\downarrow E \longrightarrow \mathbb{V}\cdot\mathfrak{X}\downarrow E;$

(4) $x\in E \longrightarrow \mathbb{V}\cdot x\downarrow E.$

By $\mathbb{V}$ we always denote the filter of neighborhoods of $0 \in \mathbb{R}$. For the meaning of $\mathfrak{X}_1 + \mathfrak{X}_2, \lambda \cdot \mathfrak{X}, \mathbb{V} \cdot \mathfrak{X}$ and $\mathbb{V} \cdot x$, see the remark following (2.4.1) above.

Conversely, if for a given vector space we say what filters converge to zero, and if the set of these filters not only satisfies the conditions (2.1.1) for x = 0, but also the above compatibility conditions (2.4.2), then we obtain, taking (2.4.1) as definition, a unique compatible pseudo-topology on E.

Looking at the induced structures studied in section 2.3 for the case of pseudo-topological vector spaces, one easily verifies that for linear maps f_i the induced pseudo-topology is also compatible. We therefore have in particular: Vector subspaces, direct products and projective limits of pseudo-topological vector spaces are also pseudo-topological vector spaces.

Suppose now that for a pseudo-topological vector space E also the following condition holds:

(2.4.3) $$\sup_{\mathfrak{X} \downarrow E} \mathfrak{X} \downarrow E.$$

By (2.4.1) we have then more generally:

$$\mathfrak{U}_x = \sup_{\mathfrak{X} \downarrow_x E} \downarrow_x E,$$

and hence

$$\mathfrak{X} \downarrow_x E \Longleftrightarrow \mathfrak{X} \leq \mathfrak{U}_x.$$

Let $U \in \mathcal{U}_x$. Then $U = x + V$, where $V \in \mathcal{U}_o$. Since by (2.4.2) we have $\mathcal{U}_o \subset \mathcal{U}_o + \mathcal{U}_o$, there exists $V' \in \mathcal{U}_o$ with $V \supset V' + V'$. Hence, with $U' = x + V' \in \mathcal{U}_x$, we have $U \supset U' + V' \supset y + V' \in \mathcal{U}_y$ for all $y \in U'$. We thus have shown: for each $U \in \mathcal{U}_x$ there exists $U' \in \mathcal{U}_x$ with $U \in \mathcal{U}_y$ for all $y \in U'$. Or it is well known that this implies that the filters $\mathcal{U}_x$ are the neighborhood filters with respect to a topology. This yields the following (cf [4]):

(2.4.4) Proposition. Condition (2.4.3) is necessary and sufficient in order that a pseudo-topological vector space is a topological vector space.

2.5. Quasi-bounded and equable filters.

On pseudo-topological vector spaces, filters $\mathcal{B}$ with the property $\mathbb{V} \cdot \mathcal{B} \downarrow E$ will be frequently used. We call them quasi-bounded filters on E. The name is motivated by the following result.

(2.5.1) Lemma. On a normed vector space E, a filter is quasi-bounded if and only if it contains a bounded set.

Proof. Let $\mathbb{V} \cdot \mathcal{B} \downarrow E$. Hence $\mathbb{V} \cdot \mathcal{B} \leq \mathcal{U}$, where $\mathcal{U}$ is the neighborhood filter of zero in E. If V denotes the unit ball in E, i.e. $V = \{x \in E \mid \|x\| \leq 1\}$, then $V \in \mathcal{U}$, hence a fortiori $V \in \mathbb{V} \cdot \mathcal{B}$.

Each set of $\mathbb{W}$ contains a closed interval $I_\delta = \{\lambda \in \mathbb{R} \mid |\lambda| \leq \delta\}$ where $\delta > 0$, and so there exists $\delta > 0$ and $B \in \mathfrak{B}$ such that $V \supset I_\delta \cdot B$. In particular, if $x \in B$, then $\|\delta \cdot x\| \leq 1$, which yields $\|x\| \leq 1/\delta$ for $x \in B$, showing that B is bounded. Conversely, suppose $\mathfrak{B}$ contains a bounded set B; so there exists $\alpha \in \mathbb{R}$ with $\|x\| < \alpha$ for all $x \in B$. If $U \in \mathfrak{U}$, choose $\varepsilon > 0$ such that $x \in U$ for $\|x\| \leq \varepsilon$. Taking $\delta = \varepsilon/\alpha$ one has $I_\delta \cdot B \subset U$, hence $U \in \mathbb{W} \cdot \mathfrak{B}$, showing that $\mathfrak{U} \geq \mathbb{W} \cdot \mathfrak{B}$ and therefore $\mathbb{W} \cdot \mathfrak{B} \downarrow E$.

A filter $\mathfrak{X}$ on a vector space is called an <u>equable filter</u> iff it has the property $\mathbb{W} \cdot \mathfrak{X} = \mathfrak{X}$. It is well known that on any topological vector space the filter of neighborhoods of zero is an equable filter.

(2.5.2) <u>Lemma</u>. A filter $\mathfrak{X}$ on a vector space is equable if and only if it satisfies the following conditions:

(1) Each $X \in \mathfrak{X}$ contains an $X' \in \mathfrak{X}$ with $I_1 \cdot X' = X'$ (*)

(2) $X \in \mathfrak{X}$, $\delta \neq 0 \Longrightarrow \delta \cdot X \in \mathfrak{X}$.

<u>Proof</u>. 1) Suppose $\mathbb{W} \cdot \mathfrak{X} = \mathfrak{X}$. Let $X \in \mathfrak{X}$. Hence $X \in \mathbb{W} \cdot \mathfrak{X}$, and thus there exist $\varepsilon > 0$ and $X_1 \in \mathfrak{X}$ with $X \supset I_\varepsilon \cdot X_1$. Taking $X' = I_\varepsilon \cdot X_1$, we have $X' \in \mathfrak{X}$ and $I_1 \cdot X' = X'$. Let further $\delta > 0$. Since, as shown, $X \supset X' = I_1 X'$, we get:

$$\delta X \supset \delta X' = \delta(I_1 X') = (\delta I_1) X' = I_\delta \cdot X' \in \mathbb{W} \cdot \mathfrak{X} = \mathfrak{X},$$

and hence $\delta X \in \mathfrak{X}$.

(*) We recall : $I_1 = \{\lambda \in \mathbb{R} \mid |\lambda| \leq 1\}$

2) Suppose, conversely, that conditions (1) and (2) hold. If $X \in \mathfrak{X}$, it follows that $X \supset X' = I_1 X'$, where $X' \in \mathfrak{X}$, and since $I_1 X' \in \mathbb{W} \cdot \mathfrak{X}$ we have $X \in \mathbb{W} \cdot \mathfrak{X}$. And if $A \in \mathbb{W} \cdot \mathfrak{X}$, then there exist $\delta > 0$ and $X_2 \in \mathfrak{X}$ with $A \supset I_\delta X_2$, hence $A \supset \delta X_2$, and since $\delta X_2 \in \mathfrak{X}$ by condition (2) we have $A \in \mathfrak{X}$, which completes the proof.

(2.5.3) <u>Lemma</u>. $\mathfrak{X}_1$ and $\mathfrak{X}_2$ equable $\Longrightarrow \mathfrak{X}_1 + \mathfrak{X}_2$ equable.

<u>Proof</u>. For arbitrary subsets X_1, X_2 of E one has $I_\delta \cdot (X_1 + X_2) \subset I_\delta X_1 + I_\delta X_2$; one easily verifies that under the hypothesis $I_1 X_1 = X_1$, $I_1 X_2 = X_2$ one gets equality: $I_\delta (X_1 + X_2) = I_\delta X_1 + I_\delta X_2$. Let now $A \in \mathbb{W} \cdot (\mathfrak{X}_1 + \mathfrak{X}_2)$. Then $A \supset I_\delta \cdot (X_1 + X_2)$, where $\delta > 0$ and $X_i \in \mathfrak{X}_i$. Since $\mathfrak{X}_i$ is equable, we can according to (2.5.2) choose X_i such that $X_i = I_1 \cdot X_i$, so that by the above equality we get: $A \supset I_\delta X_1 + I_\delta X_2$, which shows that $A \in \mathbb{W} \cdot \mathfrak{X}_1 + \mathbb{W} \cdot \mathfrak{X}_2$. We therefore have:

$$\mathbb{W} \cdot (\mathfrak{X}_1 + \mathfrak{X}_2) \geq \mathbb{W} \cdot \mathfrak{X}_1 + \mathbb{W} \cdot \mathfrak{X}_2.$$

Combining this with the converse inequality $\mathbb{W} \cdot (\mathfrak{X}_1 + \mathfrak{X}_2) \leq \mathbb{W} \cdot \mathfrak{X}_1 + \mathbb{W} \cdot \mathfrak{X}_2$, which is true for arbitrary filters, since it follows from $\lambda (x_1 + x_2) = \lambda x_1 + \lambda x_2$ by (1.5.2), we get

$$\mathbb{W} \cdot (\mathfrak{X}_1 + \mathfrak{X}_2) = \mathbb{W} \cdot \mathfrak{X}_1 + \mathbb{W} \cdot \mathfrak{X}_1 = \mathfrak{X}_1 + \mathfrak{X}_2.$$

(2.5.4) Lemma. Let $\mathfrak{X}_1$ resp. $\mathfrak{X}_2$ be filters on E_1 resp. E_2.

Then $\mathfrak{X}_1$ and $\mathfrak{X}_2$ equable $\Longrightarrow$ $\mathfrak{X}_1 \times \mathfrak{X}_2$ equable.

Proof. As before, one has for arbitrary subsets X_1 resp. X_2 of E_1 resp. E_2 :

$$I_\delta \cdot (X_1 \times X_2) \subset I_\delta X_1 \times I_\delta X_2,$$

while for subsets satisfying $I_1 X_i = X_i$ one has equality.

Therefore one gets for arbitrary filters

(2.5.5) $$\mathbb{W} \cdot (\mathfrak{X}_1 \times \mathfrak{X}_2) \leq \mathbb{W} \cdot \mathfrak{X}_1 \times \mathbb{W} \cdot \mathfrak{X}_2,$$

while for equable filters one gets as before also the converse inequality, and hence

$$\mathbb{W} \cdot (\mathfrak{X}_1 \times \mathfrak{X}_2) = \mathbb{W} \cdot \mathfrak{X}_1 \times \mathbb{W} \cdot \mathfrak{X}_2 = \mathfrak{X}_1 \times \mathfrak{X}_2.$$

(2.5.6) Lemma. Let $\mathfrak{X}_1$, $\mathfrak{X}_2$ be filters on E. Then

$\mathfrak{X}_1$ and $\mathfrak{X}_2$ equable $\Longrightarrow$ $\mathfrak{X}_1 \vee \mathfrak{X}_2$ equable.

Proof. From the set-theoretic equality

$$I_\delta \cdot (X_1 \cup X_2) = I_\delta X_1 \cup I_\delta \cdot X_2$$

one deduces, using (1.2.2), that for arbitrary filters $\mathfrak{X}_1$, $\mathfrak{X}_2$ on E :

(2.5.7) $$\mathbb{W} \cdot (\mathfrak{X}_1 \vee \mathfrak{X}_2) = \mathbb{W} \cdot \mathfrak{X}_1 \vee \mathbb{W} \cdot \mathfrak{X}_2.$$

From this, the lemma follows at once.

2.6. Equable pseudo-topological vector spaces.

A pseudo-topological vector space E is called equable iff for each $\mathfrak{X}$ with $\mathfrak{X} \downarrow E$ there exists an equable filter $\mathcal{Y} \geq \mathfrak{X}$ with $\mathcal{Y} \downarrow E$; i.e. iff

(2.6.1) $$\mathfrak{X} \downarrow E \longrightarrow \mathfrak{X} \leq \mathcal{Y} = \mathcal{W}\mathcal{Y} \downarrow E.$$

It follows from the remark preceding lemma (2.5.2), that each topological vector space is equable. However, not all pseudo-topological vector spaces are equable (*).

Given any pseudo-topological vector space E, we can introduce on $\underline{E}$ a new pseudo-topology, thus obtaining a new pseudo-topological vector space $E^{\#}$. It is defined as follows:

(2.6.2) (1) $\underline{E^{\#}} = \underline{E}$;

(2) $\mathfrak{X} \downarrow E^{\#}$ iff there exists $\mathcal{Y}$ with $\mathfrak{X} \leq \mathcal{Y} = \mathcal{W} \cdot \mathcal{Y} \downarrow E$.

One has to verify that the set of filters caracterized by the above condition (2) satisfies the conditions (2.1.1) (for x = 0) and the compatibility conditions (2.4.2). Of the conditions (2.1.1), we only verify the second one, the others being obvious. So let $\mathfrak{X}_i \downarrow E^{\#}$ for i = 1,2. Hence there exist filters $\mathcal{Y}_i$ with

$$\mathfrak{X}_i \leq \mathcal{Y}_i = \mathcal{W} \cdot \mathcal{Y}_i \downarrow E, \quad i = 1,2.$$

(*) Examples will be given in a forthcoming publication.

Using lemma (2.5.6) we get therefore

$$\mathfrak{X}_1 \vee \mathfrak{X}_2 \leq \mathfrak{Y}_1 \vee \mathfrak{Y}_2 = \mathbb{V} \cdot (\mathfrak{Y}_1 \vee \mathfrak{Y}_2).$$

Here, $\mathfrak{Y}_1 \vee \mathfrak{Y}_2 \downarrow E$ since E satisfies the conditions (2.1.1), and we see therefore that $\mathfrak{X}_1 \vee \mathfrak{X}_2 \downarrow E^*$.

We next verify the compatibility conditions.

(1) Let $\mathfrak{X}_i \downarrow E^*$ for $i = 1,2$. As before we get, this time using lemma (2.5.3) :

$$\mathfrak{X}_1 + \mathfrak{X}_2 \leq \mathfrak{Y}_1 + \mathfrak{Y}_2 = \mathbb{V} \cdot (\mathfrak{Y}_1 + \mathfrak{Y}_2),$$

which shows that $\mathfrak{X}_1 + \mathfrak{X}_2 \downarrow E^*$, since $\mathfrak{Y}_1 + \mathfrak{Y}_2 \downarrow E$.

(2) Let $\mathfrak{X} \downarrow E^*$, and $\lambda \in \mathbb{R}$. Then

$$\mathfrak{X} \leq \mathfrak{Y} = \mathbb{V} \cdot \mathfrak{Y} \downarrow E,$$

and thus

$$\lambda \mathfrak{X} \leq \lambda \mathfrak{Y} = \lambda \cdot (\mathbb{V} \mathfrak{Y}) = \mathbb{V} \cdot (\lambda \mathfrak{Y}),$$

which shows that $\lambda \mathfrak{X} \downarrow E^*$, since $\lambda \mathfrak{Y} \downarrow E$.

(3) and (4) follow immediately, since the equality $\mathbb{V} \cdot (\mathbb{V} \cdot \mathfrak{Y}) = (\mathbb{V} \cdot \mathbb{V}) \mathfrak{Y} = \mathbb{V} \cdot \mathfrak{Y}$ implies that each filter of the form $\mathbb{V} \cdot \mathfrak{Y}$ is equable. And according to the definition (2.6.2), each filter which is equable and converges to zero in E also converges to zero in E^*. This completes the verifications.

We next remark, that E^* is clearly an equable pseudo-topological vector space and that (cf.(2.3.1)) always

(2.6.3) $$E^* \leq E,$$

with equality if and only if E is equable.

(2.6.4) Lemma. $E_1^{\#} \times E_2^{\#} = (E_1 \times E_2)^{\#}$.

Proof. 1) Let $\mathfrak{X} \downarrow E_1^{\#} \times E_2^{\#}$. Hence, for i = 1,2:

$\pi_i(\mathfrak{X}) \downarrow E_i^{\#}$, which implies $\pi_i(\mathfrak{X}) \leq \mathcal{Y}_i = \mathbb{W}\,\mathcal{Y}_i \downarrow E_i$.

Using (2.3.7) and (2.3.6) we get :

$$\mathfrak{X} \leq \pi_1(\mathfrak{X}) \times \pi_2(\mathfrak{X}) \leq \mathcal{Y}_1 \times \mathcal{Y}_2 \downarrow E_1 \times E_2.$$

This shows that $\mathfrak{X} \downarrow (E_1 \times E_2)^{\#}$, since by (2.5.4) $\mathcal{Y}_1 \times \mathcal{Y}_2$ is equable.

2) Let $\mathfrak{X} \downarrow (E_1 \times E_2)^{\#}$. Hence $\mathfrak{X} \leq \mathcal{Y} = \mathbb{W}\,\mathcal{Y} \downarrow E_1 \times E_2$.

From this we get $\pi_i(\mathfrak{X}) \leq \pi_i(\mathcal{Y}) = \pi_i(\mathbb{W}\mathcal{Y}) = \mathbb{W} \cdot \pi_i(\mathcal{Y}) \downarrow E_i$,

which implies $\pi_i(\mathfrak{X}) \downarrow E_i^{\#}$ (i= 1,2). Therefore $\mathfrak{X} \downarrow E_1^{\#} \times E_2^{\#}$.

2.7. The associated locally convex topological vector space.

For any filter $\mathfrak{X}$ on a vector space, we denote by $\widehat{\mathfrak{X}}$ or $(\mathfrak{X})^{\wedge}$ the filter generated by the convex sets of $\mathfrak{X}$;these in fact form a filter basis, since the intersection of two convex sets of $\mathfrak{X}$ is again a convex set of $\mathfrak{X}$. We thus have:

(2.7.1) $A \in \widehat{\mathfrak{X}} \iff$ there exists $X \in \mathfrak{X}$, X convex, $X \subset A$.

We further define

(2.7.2) $$\mathfrak{X}^{o} = (\mathfrak{X} \vee [0])^{\wedge}.$$

From the definitions it is obvious that one has

(2.7.3) $$\mathfrak{X} \leq \widehat{\mathfrak{X}} \leq \mathfrak{X}^{o}.$$

(2.7.4) <u>Lemma.</u> A set A belongs to $\mathfrak{X}^{0}$ if and only if A contains a set $X \in \mathfrak{X}$ which is convex and satisfies $[0,1] \cdot X = X$, where $[0,1]$ is the closed unit interval of $\mathbb{R}$.

Essentially the proof of this lemma only uses the fact that if a set X is convex and contains the point 0, then it satisfies $[0,1] \cdot X = X$.

Let now E be any pseudo-topological vector space. We introduce the following filters on $\underline{E}$:

(2.7.5) $$\mathcal{U} = \sup_{\mathfrak{X} \downarrow E} \mathfrak{X} \quad ;$$

(2.7.6) $$\mathcal{V} = \hat{\mathcal{U}} = \mathcal{U}^{0} \, .$$

The equality $\hat{\mathcal{U}} = \mathcal{U}^{0}$ holds since $[0] \downarrow E$ implies $\mathcal{U} \geq [0]$ and thus $\mathcal{U} \vee [0] = \mathcal{U}$.

(2.7.7) <u>Lemma.</u> The filter $\mathcal{V}$ has the following properties:

a) $[0] \leq \mathcal{V}$;

b) $\mathbb{V} \cdot [x] \leq \mathcal{V}$ for all $x \in \underline{E}$;

c) $\lambda \cdot \mathcal{V} \leq \mathcal{V}$ for all $\lambda \in \mathbb{R}$;

d) $\mathbb{V} \cdot \mathcal{V} \leq \mathcal{V}$;

e) $\mathcal{V} + \mathcal{V} \leq \mathcal{V}$.

<u>Proof.</u> a) Since $[0] \downarrow E$, $[0] \leq \sup_{\mathfrak{X} \downarrow E} \mathfrak{X} = \mathcal{U}$; further $\mathcal{U} \leq \mathcal{V}$ by (2.7.3).

b) Same argument, using that $\mathbb{V} \cdot [x] \downarrow E$.

c) For $\lambda = 0$, this is a). For $\lambda \neq 0$ it follows, since $\mathfrak{X} \downarrow E$ if and only if $\lambda \cdot \mathfrak{X} \downarrow E$ and since V is convex if and only if $\lambda \cdot V$ is convex.

d) Let $V \in \mathfrak{V}$. Then, by (2.7.4), V contains a set $U \in \mathfrak{U}$ which is convex and satisfies $U = [0,1] \cdot U$. By c), $-\mathfrak{V} = \mathfrak{V}$, and thus $-U \in \mathfrak{V}$, since $U \in \mathfrak{V}$. Thus we have $V \supset V \cap (-V) \supset U \cap (-U) \in \mathfrak{V} \subset \mathfrak{U}$. The set $W = U \cap (-U)$ is convex and satisfies $I_1 W = W$. In fact, if $z \in I_1 \cdot W$, then $z = \lambda x$, where $|\lambda| \leq 1$ and $x \in U \cap (-U)$, i.e. $x \in U$ and $-x \in U$. Since $[0,1] \cdot U = U$, we get $\lambda x = |\lambda| \cdot (\pm x) \in U$. Thus we have $V \supset I_1 W$, which shows that $V \in \mathbb{W} \cdot \mathfrak{V}$.

e) Let $V \in \mathfrak{V}$, and choose U as before. By (c) we have also $\frac{1}{2}U \in \mathfrak{U}$; and since $\frac{1}{2}U$ is also convex it follows that $\frac{1}{2}U \in \mathfrak{V}$. Using again the convexity of U we have: $V \supset U \supset \frac{1}{2}U + \frac{1}{2}U \in \mathfrak{V} + \mathfrak{V}$, which shows that $V \in \mathfrak{V} + \mathfrak{V}$.

We define now on $\underline{E}$ a new structure, and denote $\underline{E}$ together with this new structure by E^0 :

(2.7.8) $$\mathfrak{X} \downarrow E^0 \iff \mathfrak{X} \leq \mathfrak{V}.$$

Lemma (2.7.7) immediately implies that (2.7.8) defines a compatible pseudo-topology on $\underline{E}$ (cf. (2.1.1)) and (2.4.2)). This pseudo-topology is in fact a topology, as follows from (2.4.4) or simply using the well known fact that the conditions of lemma (2.7.7) are necessary and sufficient in order that $\mathfrak{V}$ is the neighborhood-filter of zero for a unique compatible topology on $\underline{E}$ (cf. [9]). Moreover, since $\mathfrak{V}$ has, by definition, a basis consisting of convex sets, we have:

(2.7.9) Proposition. For any pseudo-topological vector space E, the space E^0 defined above is a locally convex topological vector space.

We further remark that as a consequence of (2.7.3) one has

(2.7.10) $$E \leq E^0,$$

with equality if and only if E is itself a topological locally convex vector space.

2.8. Equable continuity.

$f : E_1 \longrightarrow E_2$ being any map between pseudo-topological vector spaces, we denote by $\Delta f: E_1 \times E_1 \longrightarrow E_2$ the map defined by

(2.8.1) $$\Delta f(a,h) = f(a + h) - f(a).$$

(2.8.2) Definition. $f: E_1 \longrightarrow E_2$ is called equably continuous iff

$$\left.\begin{array}{l} \mathcal{V}\cdot\mathcal{A}\downarrow E_1 \\ \mathcal{X}\downarrow E_1 \end{array}\right\} \Longrightarrow \Delta f(\mathcal{A},\mathcal{X}) \downarrow E_2.$$

(2.8.3) Proposition. If $f: E_1 \longrightarrow E_2$ is equably continuous, then it is continuous (i.e. continuous at each point $a \in E_1$).

Proof. Let $\mathcal{X}\downarrow_a E_1$. Then, since $\mathcal{V}\cdot[a]\downarrow E_1$ and $\mathcal{X} - [a]\downarrow E_1$, we get $\Delta f([a], \mathcal{X} - [a])\downarrow E_2$. But since $\Delta f(b,x-b) = f(x)-f(b)$ we get, using (1.5.2): $\Delta f([a], \mathcal{X} - [a]) \geq f(\mathcal{X}) - f([a])$.

Therefore $f(\mathfrak{X}) - f(a) \downarrow E_2$, i.e. $f(\mathfrak{X}) \downarrow_{f(a)} E_2$, which proves the continuity of f at a.

The notion of equable continuity does not behave conveniently with respect to composition, the composite of equably continuous maps not necessarily being equably continuous. However we will have this convenient behaviour if we add a supplementary condition.

(2.8.4) Definition. A map $f : E_1 \longrightarrow E_2$ is called a quasi-bounded map iff it sends quasi-bounded filters into quasi-bounded filters, i.e. iff

$$\mathcal{V}\mathcal{A} \downarrow E_1 \longrightarrow \mathcal{V}\cdot f(\mathcal{A}) \downarrow E_2.$$

We denote by $\underline{C}_0\,(E_1;E_2)$ the space of equably continuous and quasi-bounded maps of E_1 into E_2. $\underline{C}_0\,(E_1;E_2)$ is of course closed under addition and multiplication by scalars and is therefore a vector space.

(2.8.5) Proposition. If $f \in \underline{C}_0(E_1;E_2)$ and $g \in \underline{C}_0(E_2;E_3)$, then $g \circ f \in \underline{C}_0(E_1;E_3)$.

Proof. From the definition of the operator Δ it follows that

$$(\Delta(g\circ f))(a,h) = \Delta g(f(a), \Delta f(a,h)).$$

Hence we get by (1.5.2):

$$(\Delta(g\circ f))(\mathcal{A}, \mathfrak{X}) \leq \Delta g(f(\mathcal{A}), \Delta f(\mathcal{A}, \mathfrak{X})).$$

If we assume that $\mathfrak{V}\mathcal{A} \downarrow E_1$ and $\mathfrak{X} \downarrow E_1$, it follows from the assumptions made on f and g that the filter on the right side of this inequality converges to zero in E_3, hence also

$$(\Delta(g \circ f))(\mathcal{A}, \mathfrak{X}) \downarrow E_3.$$

which proves that $g \circ f$ is equably continuous. It remains to show that $g \circ f$ is quasi-bounded; but that follows at once from the hypothesis that f and g are quasi-bounded (using of course (1.4.1)).

(2.8.6) Proposition. If E_1 is a finite dimensional vector space with its natural topology and E_2 a normed vector space, then $\underline{C}_0(E_1;E_2)$ consists exactly of the continuous maps from E_1 into E_2.

Proof. We already know ((2.8.3)) that, even for arbitrary E_1, E_2, the elements of $\underline{C}_0(E_1;E_2)$ are continuous.

Let now, conversely, $f: E_1 \longrightarrow E_2$ be continuous. Consider a quasi-bounded filter $\mathcal{A}$ on E_1. By (2.5.1), $\mathcal{A}$ contains a bounded set. Its closure, which we denote by A, is then a set of $\mathcal{A}$ which is closed and bounded. But such a set in a finite dimensional vector space is compact. f being continuous, we conclude that f(A) is compact. Hence $f(\mathcal{A})$ contains the bounded set f(A), and by (2.5.1) this implies that $f(\mathcal{A})$ is a quasi-bounded filter. We have thus shown that f is a quasi-bounded map. It remains to show that f is equably continuous. So let, as before, $\mathfrak{V}\mathcal{A} \downarrow E_1$, and further $\mathfrak{X} \downarrow E_1$. We choose again a compact set $A \in \mathfrak{A}$, and in addition a compact neighborhood V of zero in E_1. $\mathfrak{U}_i$ denotes the neighborhood filter of zero in E_i. The set $A + V$ is compact (since A and V are), and hence the continuous map f is uniformly continuous on $A + V$. This means: for every $U_2 \in \mathfrak{U}_2$ there exists $U_1 \in \mathfrak{U}_1$ such that $f(y) - f(z) \in U_2$ for all $y, z \in A + V$ with $y-z \in U_1$. We can choose U_1 sufficiently small, such that $U_1 \subset V$, and then we have in particular: $\Delta f(a,x) = f(a+x) - f(a) \in U_2$ for all $a \in A$, $x \in U_1$, i.e.: $\Delta f(A,U_1) \subset U_2$. Therefore $U_2 \in \Delta f(\mathfrak{A},\mathfrak{U}_1)$. Since $U_2 \in \mathfrak{U}_2$ was arbitrary, we have $\Delta f(\mathfrak{A},\mathfrak{U}_1) \leq \mathfrak{U}_2$. And since $\mathfrak{X} \leq \mathfrak{U}_1$, we have a fortiori $\Delta f(\mathfrak{A},\mathfrak{X}) \leq \mathfrak{U}_2$, i.e. $\Delta f(\mathfrak{A},\mathfrak{X}) \downarrow E_2$. This establishes the equable continuity of f.

(2.8.7) Proposition. If $\ell : E_1 \longrightarrow E_2$ is linear and continuous at the origin, then ℓ is quasi-bounded and equably continuous, i.e. $\ell \in \underline{C}_0(E_1;E_2)$.

Proof. By the linearity of ℓ:

$$\lambda \cdot \ell(x) = \ell(\lambda \cdot x) \quad \text{and} \quad \Delta\ell(a,h) = \ell(h),$$

we get

$$\mathbb{V} \cdot \ell(\mathfrak{A}) = \ell(\mathbb{V} \cdot \mathfrak{A}) \text{ and } \Delta\ell(\mathfrak{A},\mathfrak{X}) = \ell(\mathfrak{X}),$$

and hence the continuity at the point zero yields the two assertions.

We denote by $\underline{L}(E_1;E_2)$ the vector space formed by the linear continuous maps from E_1 to E_2. The above lemma therefore says:

$$\underline{L}(E_1;E_2) \subset \underline{C}_0(E_1;E_2)$$

By (2.8.7) and (2.8.3), as in the topological case, the continuity of a linear map at zero implies the continuity at each point. However, for bilinear (and multilinear) maps, the situation is different: continuity at the origin does not necessarily imply continuity at all points.

(2.8.8) Lemma. A bilinear map $b: E_1 \times E_2 \longrightarrow E_3$ is equably continuous if and only if it satisfies

$$(1)\quad \left.\begin{array}{l} \mathfrak{X}_1 \downarrow E_1 \\ \mathbb{V}\,\mathcal{A}_2 \downarrow E_2 \end{array}\right\} \Longrightarrow b(\mathfrak{X}_1,\mathcal{A}_2) \downarrow E_3$$

$$(2)\quad \left.\begin{array}{l} \mathbb{V}\,\mathcal{A}_1 \downarrow E_1 \\ \mathfrak{X}_2 \downarrow E_2 \end{array}\right\} \Longrightarrow b(\mathcal{A}_1,\mathfrak{X}_2) \downarrow E_3$$

Proof. We use that for a bilinear map one has

(2.8.9) $$\Delta b((a_1,a_2),(h_1,h_2)) = b(a_1,h_2) + b(h_1,a_2) + b(h_1,h_2).$$

(1) In order to prove first that the given conditions are necessary, suppose b equably continuous and let $\mathfrak{X}_1 \downarrow E_1$; $\mathbb{V}\mathcal{A}_2 \downarrow E_2$. We put $\mathcal{A} = [0] \times \mathcal{A}_2$; $\mathfrak{X} = \mathfrak{X}_1 \times [0]$. Then, using (2.5.5) and (2.3.7), we have $\mathbb{V}\cdot\mathcal{A} \downarrow E_1 \times E_2$ and $\mathfrak{X} \downarrow E_1 \times E_2$. Hence by the equable continuity of b: $\Delta b(\mathcal{A},\mathfrak{X}) \downarrow E_3$. But this is the first condition, since for our choice of $\mathcal{A}$ and $\mathfrak{X}$ one has $\Delta b(\mathcal{A},\mathfrak{X}) = b(\mathfrak{X}_1,\mathcal{A}_2)$.

The second condition is verified similarly.

(2) Let us now suppose, conversely, that the two conditions are satisfied, and let $\mathbb{V}\cdot\mathcal{A}\downarrow E_1 \times E_2$, $\mathfrak{X}\downarrow E_1 \times E_2$. Then the filters $\mathcal{A}_i = \pi_i(\mathcal{A})$, $\mathfrak{X}_i = \pi_i(\mathfrak{X})$ satisfy $\mathbb{V}\cdot\mathcal{A}_i \downarrow E_i$, $\mathfrak{X}_i \downarrow E_i$, $i = 1,2$. Using (2.3.8), (2.8.9) and (1.5.2) we get:

$$\Delta b(\mathcal{A},\mathfrak{X}) \leq \Delta b(\mathcal{A}_1, \mathcal{A}_2, \mathfrak{X}_1, \mathfrak{X}_2) \leq b(\mathcal{A}_1, \mathfrak{X}_2) + b(\mathfrak{X}_1, \mathcal{A}_2) + b(\mathfrak{X}_1, \mathfrak{X}_2).$$

By assumption each of the 3 filters on the right converges to zero on E_3. Therefore also their sum and a fortiori the left side, which proves that b is equably continuous.

(2.8.10) Lemma. If the spaces E_1, E_2 are equable, then a bilinear map $b: E_1 \times E_2 \longrightarrow E_3$ is equably continuous if and only if it is continuous at the origin.

Proof. Necessity of the condition follows from (2.8.3). In order to prove sufficiency, let $\mathfrak{X}_1 \downarrow E_1$ and $\mathbb{V}\cdot\mathcal{A}_2 \downarrow E_2$. E_1 being equable, there exists $\mathcal{Y}_1$ with $\mathfrak{X}_1 \leq \mathcal{Y}_1 = \mathbb{V}\mathcal{Y}_1 \downarrow E_1$. Thus

$$b(\mathfrak{X}_1, \mathcal{A}_2) \leq b(\mathbb{V}\mathcal{Y}_1, \mathcal{A}_2) = b(\mathcal{Y}_1, \mathbb{V}\mathcal{A}_2) = b(\mathcal{Y}_1 \times \mathbb{V}\mathcal{A}_2).$$

But by (2.3.8), $\mathcal{Y}_1 \times \mathbb{V}\mathcal{A}_2 \downarrow E_1 \times E_2$, and from the continuity of b at the origin we obtain $b(\mathfrak{X}_1, \mathcal{A}_2) \downarrow E_3$. We thus have shown that the first of the conditions of lemma (2.8.8) is satisfied; the second one is verified in the same way, and the result follows from the lemma (2.8.8).

(2.8.11) Proposition. If $b: E_1 \times E_2 \longrightarrow E_3$ is bilinear and continuous at the origin, then b is quasi-bounded.

Proof. Let $\mathcal{V}\cdot\mathcal{A} \downarrow E_1 \times E_2$. As before, $\mathcal{A}_i = \pi_i(\mathcal{A})$ satisfy $\mathcal{V}\cdot\mathcal{A}_i \downarrow E_i$, $i = 1,2$ and $\mathcal{A} \leq \mathcal{A}_1 \times \mathcal{A}_2$. Using that $\mathcal{V} = \mathcal{V}\cdot\mathcal{V}$ and $\lambda\cdot\mu b(x_1,x_2) = b(\lambda x_1, \mu x_2)$ we get

$$\mathcal{V}\, b(\mathcal{A}) \leq \mathcal{V}\cdot b(\mathcal{A}_1, \mathcal{A}_2) = \mathcal{V}\cdot\mathcal{V}\cdot b(\mathcal{A}_1, \mathcal{A}_2) = b(\mathcal{V}\mathcal{A}_1, \mathcal{V}\mathcal{A}_2),$$

and by the continuity of b at the origin, the filter on the right and hence also the one on the left converges to zero on E_3.

2.9. Continuity with respect to the associated structures.

(2.9.1) Proposition. If a linear map $\ell : E_1 \longrightarrow E_2$ is continuous, then also $\ell: E_1^* \longrightarrow E_2^*$ and $\ell : E_1^o \longrightarrow E_2^o$ are continuous.

Proof. (1) Let $\mathfrak{X} \downarrow E_1^*$. Hence $\mathfrak{X} \leq \mathcal{Y} = \mathcal{V}\mathcal{Y} \downarrow E_1$, and we get

$$\ell(\mathfrak{X}) \leq \ell(\mathcal{Y}) = \ell(\mathcal{V}\mathcal{Y}) = \mathcal{V}\cdot\ell(\mathcal{Y}) \downarrow E_2,$$

which shows that $\ell(\mathfrak{X}) \downarrow E_2^*$.

(2) Let $\mathcal{U}_i = \sup_{\mathfrak{X} \downarrow E_i} \mathfrak{X}$, $i = 1,2$. By (2.3.3) for any continuous mapping $f : E_1 \longrightarrow E_2$ with $f(0) = 0$ we have

$$f(\mathcal{U}_1) = f(\sup_{\mathfrak{X} \downarrow E_1} \mathfrak{X}) = \sup_{\mathfrak{X} \downarrow E_1} f(\mathfrak{X}) \leq \sup_{\mathcal{Y} \downarrow E_2} \mathcal{Y} = \mathcal{U}_2.$$

But by the linearity of ℓ we get also $\ell(\mathcal{U}_1^o) = (\ell(\mathcal{U}_1))^o \leq \mathcal{U}_2^o$, which ends the proof.

(2.9.2) Proposition. If $b: E_1 \times E_2 \longrightarrow E_3$ is bilinear and continuous at the origin, then $b : E_1^{\#} \times E_2^{\#} \longrightarrow E_3^{\#}$ is continuous (and hence equably continuous by (2.6.4) and (2.8.10)).

Proof. According to (2.8.10) it is sufficient to show that $b: E_1^{\#} \times E_2^{\#} \longrightarrow E_3^{\#}$ is continuous at the point (0,0). So let $\mathfrak{X}_i \downarrow E_i^{\#}$, $i = 1,2$. Hence $\mathfrak{X}_i \leq \mathfrak{Y}_i = \mathbb{W}\,\mathfrak{Y}_i \downarrow E_i$, and $b(\mathfrak{X}_1, \mathfrak{X}_2) \leq b(\mathfrak{Y}_1, \mathfrak{Y}_2) = b(\mathbb{W}\,\mathfrak{Y}_1, \mathfrak{Y}_2) = \mathbb{W}\cdot b(\mathfrak{Y}_1, \mathfrak{Y}_2) \downarrow E_3$, which shows that $b(\mathfrak{X}_1, \mathfrak{X}_2) \downarrow E_3^{\#}$.

§ 3. DIFFERENTIABILITY AND DERIVATIVES.

In this section, the definition of differentiability is given and the most elementary results of calculus are proved.

3.1. Remainders.

Let $r : E_1 \longrightarrow E_2$ be a mapping between pseudo-topological vector spaces E_1, E_2. In order to formulate the condition which will replace the classical condition of Fréchet, we associate to r a new map $\Theta r\colon \mathbb{R}\times E_1 \longrightarrow E_2$ defined by

(3.1.1) $$\Theta r(\lambda, x) = \begin{cases} 1/\lambda \cdot r(\lambda x) & \text{if } \lambda \neq 0, \\ 0 & \text{if } \lambda = 0. \end{cases}$$

(3.1.2) Definition. $r\colon E_1 \longrightarrow E_2$ is called a remainder, and we write $r \in R(E_1;E_2)$ iff $r(0) = 0$ and

$$\mathcal{V}\cdot\mathcal{B} \downarrow E_1 \Longrightarrow \Theta r(\mathcal{V},\mathcal{B}) \downarrow E_2.$$

(3.1.3) Proposition. If $r : E_1 \longrightarrow E_2$ is a remainder, then $r\colon E_1^* \longrightarrow E_2^*$ is continuous at the point zero.

Proof. Let $\mathfrak{X} \downarrow E_1^*$. Thus $\mathfrak{X} \leq \mathcal{Y} = \mathcal{V}\mathcal{Y} \downarrow E_1$. By the definition of Θr and since $r(0) = 0$ we have $r(\lambda\cdot x) = \lambda\cdot\Theta r(\lambda, x)$, and hence using (1.5.2): $r(\mathfrak{X}) \leq r(\mathcal{Y}) = r(\mathcal{V}\cdot\mathcal{Y}) \leq \mathcal{V}\cdot\Theta r(\mathcal{V},\mathcal{Y})$.
Here the filter $\mathfrak{Z} = \mathcal{V}\cdot\Theta r(\mathcal{V},\mathcal{Y})$ satisfies $\mathfrak{Z} = \mathcal{V}\cdot\mathfrak{Z} \downarrow E_2$ and therefore $r(\mathfrak{X}) \downarrow E_2^*$, which completes the proof.

As corollary we remark that for equable spaces E_1, E_2 every remainder is continuous at the origin.

(3.1.4) Proposition. $R(E_1;E_2)$ is a vector space, i.e. if $r_i \in R(E_1;E_2)$ and $\lambda_i \in \mathbb{R}$ for $i=1,2$ then $\lambda_1 r_1 + \lambda_2 r_2 \in R(E_1;E_2)$.

Proof. It is sufficient to show

(1) $r_i \in R(E_1;E_2) \Longrightarrow r_1 + r_2 \in R(E_1;E_2)$

(2) $r \in R(E_1;E_2)$, $\lambda \in R \Longrightarrow \lambda r \in R(E_1;E_2)$

(1) From the definition (3.1.1) of the operator Θ it is obvious that

$$\Theta(r_1 + r_2)\ (\lambda,x) = \Theta r_1(\lambda,x) + \Theta r_2(\lambda,x).$$

Let now $\mathbb{V}\cdot\mathfrak{B}\downarrow E_1$. Then we get, using (1.5.2):

$$\Theta(r_1 + r_2)\ (\mathbb{V},\mathfrak{B}) \leq \Theta r_1(\mathbb{V},\mathfrak{B}) + \Theta r_2(\mathbb{V},\mathfrak{B})$$

and since each term on the right converges to zero, the left side does also.

(2) is a special case of the following result.

(3.1.5) Lemma.

$$\left.\begin{array}{l} r \in R(E_1;E_2) \\ \ell \in \underline{L}(E_2;E_3) \end{array}\right\} \longrightarrow \qquad \ell\circ r \in R(E_1;E_3)$$

Proof. The linearity of ℓ implies

$$(\Theta(\ell\circ r))\ (\lambda,x) = \ell\ (\Theta r(\lambda,x)).$$

Thus

$$(\Theta(\ell \circ r))(\mathcal{V},\mathcal{B}) = \ell(\Theta r(\mathcal{V},\mathcal{B})).$$

Supposing now that $\mathcal{V}\cdot\mathcal{B}\downarrow E_1$, we have $\Theta r(\mathcal{V},\mathcal{B})\downarrow E_2$ by the assumption made on r, and hence, using the continuity of ℓ (at the origin) : $\ell(\Theta r(\mathcal{V},\mathcal{B}))\downarrow E_3$, which completes the proof.

(3.1.6) Lemma.

$$\left.\begin{array}{l} r_{12} \in R(E_1;E_2) \\ \ell \in \underline{L}(E_1;E_2) \\ r_{23} \in R(E_2;E_3) \end{array}\right\} \Longrightarrow \quad r = r_{23}\circ(r_{12}+\ell) \in R(E_1;E_3).$$

Proof. From the definition (3.1.1) of the operator Θ and the linearity of ℓ one obtains

$$\Theta r(\lambda,x) = (\Theta(r_{23}\circ(\ell + r_{12})))(\lambda,x) = \Theta r_{23}(\lambda, \ell(x) + \Theta r_{12}(\lambda,x)).$$

Thus by (1.5.2):

$$\Theta r(\mathcal{V},\mathcal{B}) \subset \Theta r_{23}(\mathcal{V},\ell(\mathcal{B}) + \Theta r_{12}(\mathcal{V},\mathcal{B})) = \Theta r_{23}(\mathcal{V},\mathcal{A}),$$

where we put $\mathcal{A} = \ell(\mathcal{B}) + \Theta r_{12}(\mathcal{V},\mathcal{B})$, which satisfies

$$\mathcal{V}\mathcal{A} \subset \mathcal{V}\cdot\ell(\mathcal{B}) + \mathcal{V}\cdot\Theta r_{12}(\mathcal{V},\mathcal{B}).$$

If we assume now that $\mathcal{V}\mathcal{B}\downarrow E_1$, we see that $\mathcal{V}\mathcal{A}\downarrow E_2$, hence $\Theta r_{12}(\mathcal{V},\mathcal{A})\downarrow E_3$, and the above inequality for $\Theta r(\mathcal{V},\mathcal{B})$ shows that also $\Theta r(\mathcal{V},\mathcal{B})\downarrow E_3$, which proves that r is a remainder.

The next result is only valid if the second space is separated.

(3.1.7) Definition (cf [4]) A pseudo-topological space E is separated iff it satisfies

$$\mathfrak{X}\downarrow_x E \text{ and } \mathfrak{X}\downarrow_y E \Longrightarrow x = y.$$

For a pseudo-topological vector space E this implies in particular $[x]\downarrow E \Longrightarrow x = 0$ (*)

(3.1.8) Lemma. If E_2 is separated, then the only remainder $r : E_1 \longrightarrow E_2$ which is linear is the zero map.

Proof. Let $x \in E_1$. By (2.4.2), $\mathbb{V}\cdot[x]\downarrow E_1$, hence $\theta r(\mathbb{V},[x])\downarrow E_2$. But since for a linear map r one has $\theta r(\lambda,x) = r(x)$, it follows that $\theta r(\mathbb{V},[x]) = r([x]) = [r(x)]$. So we have $[r(x)]\downarrow E_2$, hence $r(x) = 0$.

3.2. Differentiability at a point.

In order to make use of (3.1.8) we assume henceforth that all spaces $E_1, E_2, \ldots$ are separated. Let $f: E_1 \longrightarrow E_2$ be any map of pseudo-topological vector spaces, and $a \in E_1$.

(3.2.1) Proposition. There exists at most one $\ell \in \underline{L}(E_1;E_2)$ such that the map r defined by

$$f(a+h) = f(a) + \ell(h) + r(h)$$

is a remainder.

(*) This condition is in fact sufficient to make E separated; cf. [4].

Proof. Suppose there exist two, ℓ_1 and ℓ_2, such that the maps r_i are remainders, where

$$r_i(h) = f(a+h) - f(a) - \ell_i(h), \qquad i = 1,2.$$

Then the map

$$r = r_1 - r_2 = \ell_2 - \ell_1$$

is by (3.1.4) a remainder and is linear, hence is zero by (3.1.8), which completes the proof.

(3.2.2) Definition. If there exists a $\ell \in \underline{L}(E_1;E_2)$ such that the map r defined by

$$f(a+h) = f(a) + \ell(h) + r(h)$$

is a remainder, then the map $f\colon E_1 \longrightarrow E_2$ is said to be differentiable at the point a; and the map $\ell \in \underline{L}(E_1;E_2)$ which by (3.2.1) is uniquely determindd, is then called the derivative of f at the point a. It will be denoted as follows: $\ell = Df(a)$ or $\ell = f'(a)$.

(3.2.3) Example: A constant map $K\colon E_1 \longrightarrow E_2$ is differentiable at each point $a \in E_1$.

It will be convenient to write $Df(a).h$ instead of $(Df(a))(h)$.

If f is differentiable at the point a, the uniqueness of the derivative implies the uniqueness of the remainder r.

We use a similar notation :

(3.2.4) $r = Rf(a)$. Therefore:

$$\begin{cases} r(h) = Rf(a)(h), \text{ and} \\ \theta r(\lambda,h) = (\theta(Rf(a)))(\lambda,h) = \theta Rf(a).(\lambda,h) \end{cases}$$

(3.2.5) Proposition. If $f: E_1 \longrightarrow E_2$ is differentiable at the point a, then $f: E_1^{\#} \longrightarrow E_2^{\#}$ is continuous at the point a.

Proof. By assumption, $f(a+h) = f(a) + \ell(h) + r(h)$, where $\ell \in \underline{L}(E_1;E_2)$ and $r \in R(E_1;E_2)$. By (3.1.3), $r: E_1^{\#} \longrightarrow E_2^{\#}$ is continuous at the point zero. Further, $\ell: E_1^{\#} \longrightarrow E_2^{\#}$ is continuous according to (2.9.1). Now the result is obvious.

Remark. If in definition (3.2.2) one only requires ℓ linear, one still gets the uniqueness of the derivative, as the proof of (3.1.8) shows. However, it is essential for the theory, to impose the condition that $\ell: E_1 \longrightarrow E_2$ is continuous, since otherwise the proof of the chain rule would not work. On the other side, we see no reason to restrict the considerations to mappings $f: E_1 \longrightarrow E_2$ which are continuous; in view of (3.2.3), what we need is that $f : E_1^{\#} \longrightarrow E_2^{\#}$ is continuous. For equable and in particular for topological spaces, this distinction disappears, since then $E_i = E_i^{\#}$.

3.3 The chain rule.

(3.3.1) **Theorem.** Suppose we have maps $E_1 \xrightarrow{f} E_2 \xrightarrow{g} E_3$. Then :

$$\left.\begin{array}{l} f \text{ differentiable at } a \in E_1 \\ g \text{ differentiable at } b=f(a) \end{array}\right\} \Rightarrow \left\{\begin{array}{l} g \circ f \text{ differentiable at } a, \\ D(g\circ f)(a)=Dg(b)\circ Df(a) \end{array}\right.$$

Proof. By assumption one has

$$f(a+h) = f(a) + \ell_1(h) + r_1(h)$$

$$g(b+k) = g(b) + \ell_2(k) + r_2(k),$$

where

$$\ell_1 = Df(a) \in \underline{L}(E_1;E_2), \qquad r_1 \in R(E_1;E_2),$$

$$\ell_2 = Dg(b) \in \underline{L}(E_2;E_3), \qquad r_2 \in R(E_2;E_3).$$

Composing the mappings one gets, using the linearity of ℓ_2 :

$$(g\circ f)(a) = (g\circ f)(a) + \ell(h) + r(h), \qquad \text{where}$$

$$\ell(h) = \ell_2(\ell_1(h))$$

and $\quad r(h) = \ell_2(r_1(h)) + r_2(\ell_1(h)+r_1(h))$

Obviously, $\ell = \ell_2 \circ \ell_1 \in \underline{L}(E_1;E_3)$. Further, using (3.1.5), (3.1.6) and (3.1.4), $r \in R(E_1;E_3)$, which completes the proof.

3.4 The local caracter of the differentiability condition.

Since we consider pseudo-topological spaces E, we have to say what we mean by "local". We call E-neighborhood of a point $x \in E$ a set U with $U \in \sup_{\mathfrak{X}\downarrow_x E} \mathfrak{X} = \mathcal{U}_x$, which means: $U \in \mathfrak{X}$

for all $\mathfrak{X}$ with $\mathfrak{X} \downarrow_x E$. By (2.7.3), each neighborhood of $x \in E^o$ is an E-neighborhood of x.

A set which is an E-neighborhood of each of its points is called an E-open set. In particular, each set which is open in E^o is an E-open set.

In [4] Fischer showed that for arbitrary pseudo-topological spaces, the E-open sets are the open sets with respect to a certain topology (p. 273). In general, the filter $\mathfrak{U}_x$ is strictly finer than the neighborhood filter of x with respect to the mentioned topology. Fischer states that in the case of pseudo-topological vector spaces (and more generally for pseudo-topological groups) equality holds and he establishes this by showing that $\mathfrak{U}_o + \mathfrak{U}_o = \mathfrak{U}_o$ (cf.[4] , Satz 6, p. 294). Since it seems to us that there is a gap in his proof, we do not use the above topology (not knowing whether it is compatible), but only the associated locally convex structure introduced in 2.7.

(3.4.1) <u>Proposition</u>. Suppose that two maps $f_i : E_1 \longrightarrow E_2$ $(i = 1,2)$ coincide in an E_1-neighborhood U of the point $a \in E_1$. Then if f_1 is differentiable at the point a, f_2 is also and $f_1'(a) = f_2'(a)$.

<u>Proof</u>. By assumption we have, for all $h \in E_1$:

$$f_1(a+h) = f_1(a) + \ell_1(h) + r_1(h),$$

where $\ell_1 \in \underline{L}(E_1;E_2)$ and $r_1 \in R(E_1;E_2)$.

We define $r_2 : E_1 \longrightarrow E_2$ by

$$f_2(a+h) = f_2(a) + \ell_1(h) + r_2(h).$$

The proposition is proved, if we show that $r_2 \in R(E_1;E_2)$. Since we have $f_1(x) = f_2(x)$ for $x \in U$, we get:

$$r_2(h) = r_1(h) \quad \text{for all} \quad h \in U - a = W.$$

Here, by (2.4.1), W is an E_1-neighborhood of 0, which means

$$W \in \mathfrak{X} \quad \text{for all } \mathfrak{X} \text{ with } \mathfrak{X}\downarrow E_1.$$

Let now $\mathfrak{B}$ be a quasi-bounded filter on E_1: $\mathcal{V}\,\mathfrak{B}\downarrow E_1$. Hence $W \in \mathcal{V}\cdot\mathfrak{B}$, so there exist $V \in \mathcal{V}$ and $B \in \mathfrak{B}$ with $V.B \subset W$. One has therefore:

$$\theta r_1(\lambda,x) = \theta r_2(\lambda,x) \qquad \text{for } \lambda \in V,\ x \in B.$$

But this implies that $\theta r_1(\mathcal{V},\mathfrak{B}) = \theta r_2(\mathcal{V},\mathfrak{B})$. In fact, if $A \in \theta r_1(\mathcal{V},\mathfrak{B})$, then there exist $V_1 \in \mathcal{V}$ and $B_1 \in \mathfrak{B}$ with $A \supset \theta\, r_1(V_1,B_1) \supset \theta r_1(V_1 \cap V, B_1 \cap B) = \theta\, r_2(V_1 \cap V, B_1 \cap B)$, showing that $A \in \theta r_2(\mathcal{V},\mathfrak{B})$. Similarly for the converse. But now it is obvious that $r_1 \in R(E_1;E_2)$ implies $r_2 \in R(E_1;E_2)$.

Because of the local caracter of the differentiability condition it would be appropriate to consider maps defined on $\mathcal{E}$-open subsets of the respective pseudo-topological vector spaces E. For a map $f: A \longrightarrow E_2$, where $A \subset E_1$ is an $\mathcal{E}_1$-open set, one would introduce differentiability at a point $a \in A$ and $f'(a)$ by choosing any extension $\bar{f}$ of f to E_1, e.g. by taking $\bar{f}(x) = 0$ for $x \notin A$. However, in order to simplify the presentation, we shall continue to consider maps defined on the whole space.

§ 4. EXAMPLES AND SPECIAL CASES.

4.1. The classical case.

(4.1.1) Proposition. If E_1, E_2 are normed vector spaces, on which we consider the pseudo-topology (i.e. topology) determined by the norm, then the notions of differentiability at a point and derivative of a map $f: E_1 \longrightarrow E_2$ coincide with the classical notions in the sense of Fréchet (cf. [3]).

Proof. All we have to show is that a map $r: E_1 \longrightarrow E_2$ is a remainder (i.e. $r \in R(E_1;E_2)$, cf. definition (3.1.2)) if and only if it satisfies the classical Fréchet condition

(4.1.2) $$\lim_{x \longrightarrow 0} \frac{\|r(x)\|}{\|x\|} = 0.$$

(1) Suppose r satisfies (4.1.2), and let $\mathbb{V}\mathcal{B}\downarrow E_1$. We put

(4.1.3) $$q(x) = \begin{cases} \frac{1}{\|x\|} \cdot \|r(x)\| & \text{if } x \neq 0, \\ 0 & \text{if } x = 0. \end{cases}$$

Then, by (4.1.2), q is continuous at the point 0, and thus $q(\mathbb{V}\cdot\mathcal{B})\downarrow \mathbb{R}$, which means: $q(\mathbb{V}\cdot\mathcal{B}) \leq \mathbb{V}$.

Since one has $\|\theta r(\lambda,x)\| = \|x\| \cdot q(\lambda x)$, one gets using (1.5.2):

$$\|\theta r(\mathbb{V},\mathcal{B})\| \leq \|\mathcal{B}\| \cdot q(\mathbb{V}\cdot\mathcal{B}) = \|q(\mathbb{V}\cdot\mathcal{B})\cdot\mathcal{B}\| \leq \|\mathbb{V}\cdot\mathcal{B}\|.$$

But since $\mathbb{V}\cdot\mathcal{B}\downarrow E_1$ and the norm is continuous, $\|\mathbb{V}\cdot\mathcal{B}\|\downarrow\mathbb{R}$, hence also $\|\theta r(\mathbb{V},\mathcal{B})\|\downarrow\mathbb{R}$, which yields, by the definition of the pseudo-topology induced by the norm: $\theta r(\mathbb{V},\mathcal{B})\downarrow E_1$. This shows that $r \in R(E_1;E_2)$.

(2) Suppose conversely that $r \in R(E_1;E_2)$ and let q be as before. It is sufficient to show that q is continuous at the point 0. So let $\mathfrak{X} \downarrow E_1$; we shall show that then $q(\mathfrak{X}) \downarrow \mathbb{R}$. We still introduce the map $s: E_1 \longrightarrow E_1$ by

$$s(x) = \begin{cases} \frac{1}{\|x\|} \cdot x & \text{for } x \neq 0, \\ 0 & \text{for } x = 0. \end{cases}$$

Since for all $x \in E_1$, $\|s(x)\| \leq 1$, $s(\mathfrak{X})$ certainly contains a bounded set, hence by (2.5.1) is a quasi-bounded filter on E_1. We further have $\|\mathfrak{X}\| \leq \mathbb{V}$ and $q(x) = \|\theta r(\|x\|, s(x))\|$. So we get, using (1.5.2):

$$q(\mathfrak{X}) \leq \|\theta r(\|\mathfrak{X}\|, s(\mathfrak{X}))\| \leq \|\theta r(\mathbb{V}, s(\mathfrak{X})\| \downarrow \mathbb{R}.$$

4.2. Linear and bilinear maps.

(4.2.1) Proposition. A linear and continuous map $f: E_1 \longrightarrow E_2$ is differentiable at each point $a \in E_1$ and $f'(a) = f$.

The proof is obvious: taking $\ell = f$ and $r = 0$ one has $f(a+h) = f(a) + \ell(h) + r(h)$, and $\ell \in \underline{L}(E_1;E_2)$, $r \in R(E_1;E_2)$.

(4.2.2) Lemma. Let $b: E_1 \times E_2 \longrightarrow E_3$ be bilinear and continuous at the point zero. Then $b \in R(E_1 \times E_2; E_3)$.

Proof. One has for all $\lambda \in \mathbb{R}$, $(x_1,x_2) \in E_1 \times E_2$:

$$\theta b(\lambda,(x_1,x_2)) = \lambda \cdot b(x_1,x_2).$$

Hence $\theta b(\mathbb{V},\mathfrak{B}) = \mathbb{V} \cdot b(\mathfrak{B})$.

Further, since $\lambda . \mu . b(x_1,x_2) = b(\lambda x_1, \mu x_2)$, one has

$$\mathbb{V}.b(\mathfrak{B}_1, \mathfrak{B}_2) = \mathbb{V}. \mathbb{V}.b(\mathfrak{B}_1, \mathfrak{B}_2) = b(\mathbb{V}\mathfrak{B}_1, \mathbb{V}\mathfrak{B}_2).$$

Suppose now that $\mathcal{V}\cdot\mathcal{B}\downarrow E_1 \times E_2$. One has $\mathcal{B} \leq \Pi_1(\mathcal{B}) \times \Pi_2(\mathcal{B})$, where $\mathcal{B}_i = \Pi_i(\mathcal{B})$ satisfies $\mathcal{V}\mathcal{B}_i \downarrow E_i$, $i = 1,2$. Hence

$$\Theta b(\mathcal{V},\mathcal{B}) = \mathcal{V}.b(\mathcal{B}) \leq \mathcal{V}.b(\mathcal{B}_1,\mathcal{B}_2) = b(\mathcal{V}\mathcal{B}_1,\mathcal{V}\mathcal{B}_2)\downarrow E_3.$$

(4.2.3) <u>Proposition</u>. Let $b: E_1 \times E_2 \longrightarrow E_3$ be bilinear and continuous. Then b is differentiable at each point $a = (a_1,a_2) \in E_1 \times E_2$, and

$$b'(a_1,a_2)(h_1,h_2) = b(h_1,a_2) + b(a_1,h_2).$$

<u>Proof</u>. One has, for $h = (h_1,h_2) \in E_1 \times E_2$:

$$b(a+h) = b(a_1+h_1,a_2+h_2) = b(a_1,a_2) + b(h_1,a_2) + b(a_1,h_2) + b(h_1,h_2).$$

So we have, with $\ell(h) = \ell(h_1,h_2) = b(h_1,a_2) + b(a_1,h_2)$ and $r(h) = r(h_1,h_2) = b(h_1,h_2)$:

$$b(a+h) = b(a) + \ell(h) + r(h).$$

ℓ is obviously linear, since b is bilinear, and also continuous, since b is continuous. Thus $\ell \in \underline{L}(E_1 \times E_2;E_3)$. And $r \in R(E_1 \times E_2;E_3)$ by the preceding lemma. This completes the proof.

<u>4.3. The special case $f: \mathbb{R} \longrightarrow E$.</u>

(4.3.1) <u>Proposition</u>. If $f: \mathbb{R} \longrightarrow E$ is differentiable at the point $\alpha \in \mathbb{R}$, then the following limit, which we denote by $f^\bullet(\alpha)$, exists:

(4.3.2) $$f^\bullet(\alpha) = \lim_{\xi \to 0} \frac{f(\alpha+\xi)-f(\alpha)}{\xi}.$$

Further one then has:

$$\left(f'(\alpha)\right)(\xi) = \xi \cdot f^\bullet(\alpha) \quad ; \quad f^\bullet(\alpha) = \left(f'(\alpha)\right)(1).$$

Proof. We have $f(\alpha+\xi) = f(\alpha) + \ell(\xi) + r(\xi)$, where $\ell \in \underline{L}(\mathbb{R};E)$ and $r \in R(\mathbb{R};E)$. We put $\ell(1) = f'(\alpha).1 = a$ and define $q: \mathbb{R} \longrightarrow E$ by

$$q(\xi) = \begin{cases} \dfrac{f(\alpha+\xi) - f(\alpha)}{\xi} & \text{if } \xi \neq 0, \\ a & \text{if } \xi = 0. \end{cases}$$

One has

$$\ell(\xi) = \ell(\xi.1) = \xi.\ell(1) = \xi.a.$$

Hence

$$r(\xi) = f(\alpha+\xi) - f(\alpha) - \xi.a,$$

$$\theta r(\lambda,\xi) = \frac{f(\alpha+\lambda\xi) - f(\alpha)}{\lambda} - \xi.a,$$

$$\theta r(\lambda,1) = q(\lambda) - a.$$

Since $\mathbb{V}.1 \downarrow \mathbb{R}$ we have (Definition 3.1.2) $\theta r(\mathbb{V},1) \downarrow E$, hence $q(\mathbb{V}) - a \downarrow E$ resp. $q(\mathbb{V}) \downarrow_a E$, and that means exactly that $\lim_{\xi \to 0} q(\xi) = a$; the proof is complete.

In the classical case, the converse of this proposition holds. However it seems, that in the general case, our differentiability condition is a little bit stronger. Again some question of equability comes in.

(4.3.3) Proposition. If the scalar multiplication $\mathbb{R} \times E \longrightarrow E$ of E is equably continuous (thus by (2.8.10) in particular if E is equable or even topological)(*), then the existence of the differential quotient (4.3.2) is sufficient for the differentiability of $f: \mathbb{R} \longrightarrow E$ at the point α .

(*) More generally if E is admissible (see appendix (1)).

Proof. Let $a = \lim_{\xi \to 0} \frac{f(\alpha+\xi)-f(\alpha)}{\xi}$, q being defined as before; we can write therefore:

$$q(\mathcal{V})-a \downarrow E.$$

We define $\ell : \mathbb{R} \longrightarrow E$ by : $\ell(\xi) = \xi.a$, and $r : \mathbb{R} \longrightarrow E$ by $r(\xi) = f(\alpha+\xi) - f(\alpha) - \ell(\xi)$. Obviously $\ell \in \underline{L}(\mathbb{R};E)$, and it only remains to prove that $r \in R(\mathbb{R};E)$. We have, according to the above definition of r:

$$\theta r(\lambda,\xi) = \xi.(q(\lambda\xi)-a).$$

Thus we get, if $\mathcal{V}\mathcal{B} \downarrow \mathbb{R}$, i.e. if $\mathcal{V}\mathcal{B} \leq \mathcal{V}$, using (1.5.2):

$$\theta r(\mathcal{V},\mathcal{B}) \leq \mathcal{B}.(q(\mathcal{V}\mathcal{B})-a) \leq \mathcal{B}.(q(\mathcal{V})-a).$$

On the right, we have the product of a quasi-bounded filter on $\mathbb{R}$ with a zero-converging filter on E, which converges to zero under the hypothesis that scalar multiplication is equably continuous.

4.4. Differentiable mappings into a direct product.

(4.4.1) Proposition. Let $f_i : F_i \longrightarrow E_i$, $i \in I$, be a family of mappings of pseudo-topological vector spaces. Then (cf (1.3.1)) $\times_{i \in I} f_i : \times_{i \in I} F_i \longrightarrow \times_{i \in I} E_i$ is differentiable at the point $\{x_i\}_{i \in I}$ if and only if $f_i : F_i \longrightarrow E_i$ is differentiable at the point x_i for all $i \in I$, and then $(\times_{i \in I} f_i)'(\{x_i\}_{i \in I}) = \times_{i \in I} f_i'(x_i)$.

In particular:

$$(f_1 \times f_2)'(x_1,x_2) = (f_1'(x_1),\ f_2'(x_2))$$

The proof is a combination of the following two lemmas.

(4.4.2) Lemma. Let $\ell_i : F_i \longrightarrow E_i$, $i \in I$, be maps. Then

$\underset{i\in I}{\times} \ell_i \in L(\underset{i\in I}{\times} F_i;\ \underset{i\in I}{\times} E_i) \Longleftrightarrow \ell_i \in L(F_i;E_i)$ for all $i\in I$.

Proof. One verifies separately that $\underset{i\in I}{\times} \ell_i$ is linear if and only if each ℓ_i is linear and that $\underset{i\in I}{\times} \ell_i$ is continuous if and only if each ℓ_i is continuous.

(4.4.3) Lemma. Let $r_i : F_i \longrightarrow E_i$, $i\in I$, be maps. Then

$\underset{i\in I}{\times} r_i \in R(\underset{i\in I}{\times} F_i;\underset{i\in I}{\times} E_i) \Longleftrightarrow r_i \in R(F_i;E_i)$ for all $i\in I$.

Proof. Let us denote, for $j\in I$, by $\pi_j : \underset{i\in I}{\times} E_i \longrightarrow E_j$ and $w_j : \underset{i\in I}{\times} F_i \longrightarrow F_j$ the projections, and put $r = \underset{i\in I}{\times} r_i$.

One has

(4.4.4) $$r_j \circ w_j = \pi_j \circ r.$$

a) Let $r_j \in R(F_j;E_j)$ for all $j\in I$. From (3.1.6) and (4.4.4) it follows that $\pi_j \circ r \in R(\underset{i\in I}{\times} F_i;E_j)$ for all $j\in I$. Suppose now that $\mathcal{V}\mathcal{B} \downarrow \underset{i\in I}{\times} F_i$. Then $\theta(\pi_j \circ r)(\mathcal{V},\mathcal{B}) \downarrow E_j$ for all $j\in I$. But π_j being linear we have $\theta(\pi_j \circ r) = \pi_j \circ \theta r$. Hence $\pi_j(\theta r(\mathcal{V},\mathcal{B})) \downarrow E_j$ for all $j\in I$, which yields $\theta r(\mathcal{V},\mathcal{B}) \downarrow \underset{i\in I}{\times} E_i$. This shows that $r \in R(\underset{i\in I}{\times} F_i;\underset{i\in I}{\times} E_i)$.

b) Let $r \in R(\bigtimes_{i\in I} F_i; \bigtimes_{i\in I} E_i)$. Then, by (3.1.5) and (4.4.4) we get $r_j \circ w_j \in R(\bigtimes_{i\in I} F_i; E_j)$ for all $j \in I$. Denoting by $\varphi_j\colon F_j \longrightarrow \bigtimes_{i\in I} F_i$ the map caracterized by the conditions $w_k \circ \varphi_j = 0$ for $k \neq j$, $w_j \circ \varphi_j$ = identity, we now conclude from the continuity and linearity of φ_j by (3.1.6) that $r_j = r_j \circ (w_j \circ \varphi_j) = (r_j \circ w_j) \circ \varphi_j$ belongs to $R(F_j; E_j)$.

(4.4.5) <u>Proposition</u>. Let $f_i : E \longrightarrow E_i$, $i \in I$, be a family of mappings of pseudo-topological vector spaces. Then (cf.(1.3.2)) $\prod_{i\in I} f_i : E \longrightarrow \bigtimes_{i\in I} E_i$ is differentiable at the point $x \in E$ if and only if $f_i : E \longrightarrow E_i$ is differentiable at the point x for all $i \in I$, and then:

$$\Big(\prod_{i\in I} f_i\Big)'(x) = \prod_{i\in I} f_i'(x).$$

In particular:

$$[f_1, f_2]'(x) = [f_1'(x), f_2'(x)] .$$

<u>Proof</u>. a) Let all the maps f_i be differentiable at x. Then by the preceding proposition $\bigtimes_{i\in I} f_i$ is differentiable at $d(x)$, where d is the diagonal map of E into its I-fold direct product E^I. Further, the map $d\colon E \longrightarrow E^I$, being linear and continuous, is differentiable by (4.2.1). Now the differentiability of $\prod_{i\in I} f_i$ and also the given formula for its derivative follow from the chain rule, (1.3.3) and (4.2.1).

b) Let $\prod_{i \in I} f_i$ be differentiable. Since $f_j = \pi_j \circ \prod_{i \in I} f_i$, where the projection map π_j is linear and continuous, the differentiability of f_j follows again from the chain rule and (4.2.1).

§ 5. FUNDAMENTAL THEOREM OF CALCULUS.

The value of a vector-valued function will be estimated by means of its derivative. Since no norm is available, estimation is formulated by means of convex sets, which is advantageous also in the normed case.

5.1. Formulation and proof of the main theorem.

Suppose there are two maps given: $f:[\alpha,\beta] \longrightarrow E$ and $\varphi:[\alpha,\beta] \longrightarrow \mathbb{R}$, E being a pseudo-topological vector space, such that the following conditions are satisfied

(5.1.1) E^0 separated;

(5.1.2) $f: [\alpha,\beta] \longrightarrow E^0$ and $\varphi: [\alpha,\beta] \longrightarrow \mathbb{R}$ continuous;

(5.1.3) B a closed and convex subset of E^0;

(5.1.4) for almost all (*) $t \in [\alpha,\beta]$, f and φ are differentiable at t and satisfy

$$f'(t) \in \varphi'(t).B;$$

(5.1.5) $s < t \Longrightarrow \varphi(s) \leq \varphi(t)$.

* Throughout this paper, "almost all" is never used in the sense of measure theory, but always means: "all with at most a denumerable infinity of exceptions".

(5.1.6) Theorem. Under the above hypothesis one has

$$f(\beta) - f(\alpha) \in (\varphi(\beta) - \varphi(\alpha)).B.$$

Proof. Part 1. In this part, we reduce the general case to the following special case :

(5.1.7) $\alpha = 0$; $\varphi(0) = 0$; $f(0) = 0$; $0 \in B$.

In order to do this, we choose a fixed point $p \in B$ and we define $\alpha_1, \beta_1, \varphi_1, f_1, B_1$ as follows:

$$\alpha_1 = 0 ; \quad \beta_1 = \beta - \alpha;$$

$$\varphi_1(t) = \varphi(t+\alpha) - \varphi(\alpha) \qquad \text{for } t \in [\alpha_1, \beta_1] ;$$

$$f_1(t) = f(t+\alpha) - f(\alpha) - \varphi_1(t).p \qquad \text{for } t \in [\alpha_1, \beta_1] ;$$

$$B_1 = B - p.$$

One easily verifies that the validity of the conditions (5.1.1) to (5.1.5) for $\alpha, \beta, \varphi, f, B$ implies their validity for $\alpha_1, \beta_1, \varphi_1, f_1, B_1$. In fact, let us check (5.1.4), the others being obvious. By the chain rule, we have:

$$\varphi^{\cdot}_1(t) = \varphi^{\cdot}(t+\alpha) ;$$

$$f^{\cdot}_1(t) = f^{\cdot}(t+\alpha) - \varphi^{\cdot}_1(t).p = f^{\cdot}(t+\alpha) - \varphi^{\cdot}(t+\alpha).p.$$

Hence one concludes from $f^{\cdot}(t) \in \varphi^{\cdot}(t).B$ for $t \in [\alpha, \beta]$:

$$f^{\cdot}_1(t) \in \varphi^{\cdot}(t+\alpha).B - \varphi^{\cdot}(t+\alpha).p = \varphi^{\cdot}_1(t).B_1 \quad \text{for } t \in [\alpha_1, \beta_1] .$$

Suppose now the theorem holds for the special case.

Then we have $f_1(\beta_1) \in \varphi_1(\beta_1).B_1$. But this, using the definitions of $\beta_1, \varphi_1, f_1, B_1$, yields immediately:

$$f(\beta) - f(\alpha) - (\varphi(\beta) - \varphi(\alpha)).p \in (\varphi(\beta) - \varphi(\alpha)).(B-p),$$

which is equivalent to the assertion of the theorem. Having reduced the general case to the special case (5.1.7), we henceforth assume that conditions (5.1.7) hold.

Part 2. We suppose in this part of the proof that B is, in E^0, a neighborhood of zero, and we show that then one has for all $\varepsilon > 0$:

(5.1.8) $$f(\beta) \in (\varphi(\beta) + \varepsilon.\beta + \varepsilon).B.$$

Let $\rho_1, \rho_2, \rho_3, \dots$ be an enumeration of the points where possibly (5.1.4) does not hold (*), and let us define an auxiliary function $\chi: [\alpha,\beta] \longrightarrow \mathbb{R}$ by

(5.1.9) $$\chi(s) = \varphi(s) + \varepsilon.s + \varepsilon.\sum_{\rho_n < s} (\tfrac{1}{2})^n.$$

Let $\quad I = \{t \in [0,\beta] \mid \quad f(s) \in \chi(s).B \quad \text{for } 0 \leq s < t\}$

Obviously one has : $0 \in I$. Let $\gamma = \sup I$. If $0 \leq t < \gamma$, there exists $t_1 \in I$ with $t < t_1 \leq \gamma$; hence $f(s) \in \chi(s).B$ for $0 \leq s < t_1$, thus in particular for $0 \leq s \leq t$, which shows that $t \in I$ and that $f(t) \in \chi(t).B$. But this shows that also $\gamma \in I$, and we have therefore: $I = [0,\gamma]$.

---

* We can always take α and β among the points ρ_n; then we do not have to bother what differentiability in the endpoints α , β would mean (cf.3.4).

We claim now that $f(\gamma) \in \chi(\gamma).B$. If $\gamma = 0$, nothing has to be proved. If $\gamma > 0$, we use that

$$\frac{f(t)}{\chi(t)} \in B \qquad \text{for } 0 < t < \gamma.$$

Since $f: [\alpha, \beta] \longrightarrow E^{0}$ is continuous and χ continuous from the left, it follows, using that B is closed in E^{0}, that

$$\lim_{\substack{t \to \gamma \\ t < \gamma}} \frac{f(t)}{\chi(t)} = \frac{f(\gamma)}{\chi(\gamma)} \in B,$$

which means

(5.1.10) $$f(\gamma) \in \chi(\gamma).B.$$

Since $I = [0, \gamma] \subset [0, \beta]$, we have of course

(5.1.11) $$\gamma \leq \beta.$$

In order to show that here equality holds, we proceed indirectly : we assume that $\gamma < \beta$. Then either $\gamma \neq \rho_n$ for all n or $\gamma = \rho_m$. We show that none of these cases is possible.

Case 1: $\gamma < \beta$, $\gamma \neq \rho_n$ for all n. This means that γ is not an exceptional point: f and φ are differentiable at γ and (5.1.4) holds for $t = \gamma$. Thus we get

$f(\gamma + h) = f(\gamma) + h.f'(\gamma) + r_1(h)$, where $r_1 \in R(\mathbb{R}; E^{0})$;

$\varphi(\gamma + h) = \varphi(\gamma) + h.\varphi'(\gamma) + r_2(h)$, where $r_2 \in R(\mathbb{R}; \mathbb{R})$;

$f'(\gamma) \in \varphi'(\gamma).B.$

Since $I_1 = [-1,1]$ certainly satisfies $\mathbb{W}.I_1 \downarrow \mathbb{R}$ we conclude: $\Theta r_1(\mathbb{W}, [I_1]) \downarrow E^n$, which means $\Theta r_1(\mathbb{W}, [I_1]) \leq \mathcal{V}$, $\mathcal{V}$ being the neighborhood filter of zero in E^0. Since in this part 2 of the proof we assume that $B \in \mathcal{V}$, we have also $\varepsilon/2.B \in \mathcal{V} \subset \Theta r_1(\mathbb{W}, [I_1])$, and hence there exists $\delta_1 > 0$ such that $\Theta r_1(I_{\delta_1}, I_1) \subset \varepsilon/2\, B$, and thus (cf (3.1.1)):

$$r_1(\lambda) \in \varepsilon/2.\lambda.B \qquad \text{for } |\lambda| \leq \delta_1$$

In the same way we conclude: there exists $\delta_2 > 0$ such that

$$r_2(\lambda) \in \varepsilon/2.\lambda.I_1 \qquad \text{for } |\lambda| \leq \delta_2, \quad \text{i.e.}$$

$$|r_2(\lambda)| \leq \varepsilon/2.|\lambda| \qquad \text{for } |\lambda| \leq \delta_2.$$

Let now $\delta = \mathrm{Min}(\delta_1, \delta_2, \beta - \gamma)$. Then we have for $0 \leq H \leq \delta$:

$$\begin{aligned} f(\gamma+h) &= f(\gamma) + h.f^\cdot(\gamma) + r_1(h) \\ &\in \varkappa(\gamma).B + h.\varphi^\cdot(\gamma).B + h.\ \varepsilon/2.B \\ &= \varkappa(\gamma).B + (\varphi(\gamma+h) - \varphi(\gamma) - r_2(h)).B + h.\ \varepsilon/2.B. \end{aligned}$$

Here, the coefficients $\varkappa(\gamma)$, $\varphi(\gamma+h) - \varphi(\gamma) - r_2(h) = h.\varphi^\cdot(\gamma)$ and $h.\ \varepsilon/2$ are, using (5.1.5), all non-negative. Using that for a convex set B and non-negative coefficients μ, ν, σ one has $\mu.B + \nu.B + \sigma.B \subset (\mu+\nu+\sigma).B$, we get therefore:

$$\begin{aligned} f(\gamma+h) &\in \left(\varkappa(\gamma) + \varphi(\gamma+h) - \varphi(\gamma) - r_2(h) + h.\ \varepsilon/2\right).B \\ &= \Big(\varphi(\gamma+h) + \varepsilon.\sum_{\rho_n < \gamma} (\tfrac{1}{2})^n + \varepsilon.\gamma - r_2(h) + h\ \varepsilon/2\Big).B \\ &\subset \Big(\varphi(\gamma+h) + \varepsilon.\sum_{\rho_n < \gamma+h} (\tfrac{1}{2})^n + \varepsilon.(\gamma+h)\Big).B \\ &= \varkappa(\gamma+h).B. \end{aligned}$$

Thus one would have $f(s) \in \chi(s).B$ for $0 \leq s \leq \gamma$ and $\gamma \leq s \leq \gamma + \delta$, hence $\gamma + \delta \in I$, which contradicts the definition of γ as supremum of I. This shows that case 1 is not possible.

Case 2: $\gamma < \beta, \gamma = \rho_m$. Since f is continuous at the point ρ_m, there exists $\delta_1 > 0$ with

$$f(\xi) - f(\gamma) \in \frac{\varepsilon}{2} \cdot \frac{1}{2^m} .B \qquad \text{for } |\xi - \gamma| \leq \delta_1 .$$

Analogously, since φ is continuous and monotonic: there exists $\delta_2 > 0$ with

$$\varphi(\xi) - \varphi(\gamma) \leq \varepsilon/2 . \frac{1}{2^m} \qquad \text{for } \gamma < \xi \leq \gamma + \delta_2 .$$

Putting again $\delta = \text{Min}(\delta_1, \delta_2, \beta - \gamma)$, we have for $\gamma < \xi \leq \gamma + \delta$:

$$\begin{aligned} f(\xi) &= (f(\xi) - f(\gamma)) + f(\gamma) \\ &\in \frac{\varepsilon}{2} \cdot \frac{1}{2^m} B + (\varphi(\gamma) + \varepsilon\gamma + \varepsilon \sum_{\rho_n < \gamma} \frac{1}{2^n}).B \\ &\subset \frac{\varepsilon}{2} \frac{1}{2^m} B + (\varphi(\xi) + \varepsilon\xi + \varepsilon \sum_{\rho_n < \gamma} \frac{1}{2^n}).B \\ &\subset (\varphi(\xi) + \varepsilon \cdot \xi + \varepsilon . \sum_{\rho_n \leq \gamma} \frac{1}{2^n}).B \qquad \subset \chi(\xi).B. \end{aligned}$$

As before in case 1, it would follow from this that $\gamma + \delta \in I$, in contradiction with the definition of γ. Hence case 2 is not possible either. The consequence of this indirect argument is that the assumption $\gamma < \beta$ is impossible, and by (5.1.11) the equality $\gamma = \beta$ must hold. But then, (5.1.10) yields exactly (5.1.8), i.e. what we wanted to show in this part of the proof.

Part 3. We still suppose, as in part 2, that B is a neighborhood of zero in E^0, and we show, that then the conclusion of the theorem holds, namely

(5.1.12) $$f(\beta) \in \varphi(\beta).B.$$

If $\varphi(\beta) \neq 0$, this follows immediately, using (5.1.8) and the fact that B is closed, by passing to the limit for $\varepsilon \longrightarrow 0$:

$$\frac{f(\beta)}{\varphi(\beta)} = \lim_{\substack{\varepsilon \to 0 \\ \varepsilon > 0}} \frac{f(\beta)}{\varphi(\beta) + \varepsilon\beta + \varepsilon} \in B.$$

So let us consider the case $\varphi(\beta) = 0$. Since $\varphi(0) = 0$, the monotonicity of φ implies that then $\varphi(t) = 0$ for all $t \in [0,\beta]$. Hypothesis (5.1.4) then yields: $f'(t) = 0$ for almost all $t \in [0,\beta]$. Thus for any closed, convex neighborhood C of the origin in E^0, the five points of the hypothesis of the theorem are satisfied (with B replaced by C), and according to Part 2 which is already proved one has $f(\beta) \in (\varepsilon\beta + \varepsilon).C$, or in particular

$$\frac{f(\beta)}{\beta + 1} \in C.$$

This being true for any closed, convex neighborhood of zero in E^0, it follows from (5.1.1) that $\frac{f(\beta)}{\beta+1} = 0$, and since $0 \in B$, we have again (5.1.12).

<u>Part. 4</u> Having reduced in Part 1 the general case to the special case (5.1.7), we still assume that (5.1.7) holds. It only remains to get rid of the additional hypothesis made in Part 2 and Part 3, namely that B is a neighborhood of zero in E^0. So let now B be any closed convex set containing zero. If $\varphi(\beta) = 0$, we already know that $f(\beta) = 0$ and thus certainly $f(\beta) \in \varphi(\beta).B$.

In order to show that this also holds in the case $\varphi(\beta) \neq 0$, we assume (indirect proof) the contrary:

(5.1.13) $$z = \frac{f(\beta)}{\varphi(\beta)} \notin B.$$

Since B is closed (always in E^o), we can choose a neighborhood U of z with $U \cap B = \emptyset$. By the continuity of the addition, since $o - o + z = z \in U$, we can choose a neighborhood V of 0 such that $V - V + z \subset U$. The topology of E^o being locally convex, we can choose a V which is convex.

Now, $(z + V) \cap (B + V) = \emptyset$. In fact, if this intersection were not empty, there would exist points $v_1, v_2 \in V$ and $b \in B$ with $z + v_1 = b + v_2$, thus $b = v_1 - v_2 + z \in V - V + z \subset U$, which contradicts $U \cap B = \emptyset$.

Hence $z + V$ is a neighborhood of z which is disjoint from $B + V$, and therefore $z \notin \overline{B + V} = B^*$. The set B^* is a closed, convex neighborhood of zero in E^o, and since $B^* \supset B$, the hypothesis of the theorem are a fortiori satisfied for B^* instead of B, and according to Part 3 we conclude that $f(\beta) \in \varphi(\beta).B^*$, or equivalently: $z \in B^*$, in contradiction with $z \notin B^*$. This implies that assumption (5.1.13) is impossible, which concludes the proof of the fundamental theorem.

5.2. Remarks and special cases.

(5.2.1) *Proposition*. The fundamental theorem (5.1.6) is also valid if the hypothesis that $f: [\alpha,\beta] \longrightarrow E^0$ is continuous and differentiable at almost all points $t \in [\alpha,\beta]$ is replaced by the hypothesis that $f: [\alpha,\beta] \longrightarrow E$ is continuous and differentiable at almost all points.

Proof. Since by (2.7.10) $E \leqslant E^0$, the identity map $i: E \longrightarrow E^0$ is continuous, thus (being linear) also differentiable. Hence the statement follows from the fact that the composite of continuous maps is continuous and the chain rule. We further remark that $f'(t)$ is then the same element of $\underline{E} = \underline{E}^0$ whether we consider $f: [\alpha,\beta] \longrightarrow E$ or $f: [\alpha,\beta] \longrightarrow E^0$.

(5.2.2) *Corollary*. Let $\varphi: [\alpha,\beta] \longrightarrow \mathbb{R}$ be continuous and satisfy for almost all $t \in [\alpha,\beta]$ the inequality $\varphi'(t) \geqslant 0$. Then φ is monotonic: $s < t \Longrightarrow \varphi(s) \leqslant \varphi(t)$.

Proof. We put $B = \mathbb{R}^+ = \{x \in \mathbb{R} \mid x \geqslant 0\}$, and define an auxiliary function $\psi: [\alpha,\beta] \longrightarrow \mathbb{R}$ by $\psi(t) = t$. Then the main theorem can be applied with functions φ, ψ in the place of f, φ, and choosing any interval $[s,t] \subset [\alpha,\beta]$, noting that B is convex and closed and ψ is monotonic. It yields:

$$\varphi(t) - \varphi(s) \in (t-s).B,$$

which implies the result, since $(t-s).B = B$.

(5.2.3) Remark. For normed vector spaces E, the "mean value theorem" in its modern formulation (cf. (8.5.1) in [3]) is a special case of theorem (5.1.6). One simply chooses B to be the closed unit ball of E. We have then $E = E^0$, and (5.1.1) and (5.1.3) are satisfied. Hypothesis (5.1.2) and (5.1.4) are maintained, the condition of (5.1.4) reducing now to the inequality $\|f^{\bullet}(t)\| \leq \varphi^{\bullet}(t)$. From (5.2.2) follows, that now also hypothesis (5.1.5) becomes superfluous. And finally, the conclusion of the theorem reduces also to an inequality: $\|f(\beta) - f(\alpha)\| \leq \varphi(\beta) - \varphi(\alpha)$. So we are exactly left with the so-called mean value theorem for normed vector spaces.

However, theorem (5.1.6) is also in the case of normed vector spaces more interesting then the classical formulation which uses the norm (i.e. taking as B the unit ball), since it not only says that if for a motion of a point in E the velocity is not too big, then the point does not stray too far, but also that if the velocity is "big" in a certain sense, the point is much displaced. However, the condition that the velocity has to be "big" cannot be expressed in terms of a norm-inequality of the form $\|f^{\bullet}(t)\| \geq \ldots$, since this would involve a non-convex set. The velocity has to be "big" in the sense that $f^{\bullet}(t)$ lies in a convex set B (which does not have to contain the origin), or more generally in $\varphi^{\bullet}(t).B$.

As an example, let us consider a continuous function $\varphi: [\alpha, \beta] \longrightarrow \mathbb{R}$, differentiable at almost all points of $[\alpha, \beta]$, and suppose that $m \leq \varphi'(t) \leq M$ for almost all $t \in [\alpha, \beta]$. We can apply the theorem, taking as convex set B the interval $[m, M]$ and as comparative function the identity function $\psi(t) = t$, and we get immediately: $m(\beta - \alpha) \leq \varphi(\beta) - \varphi(\alpha) \leq M(\beta - \alpha)$.

Examples in spaces of dimension greater than 1 are of course more interesting.

5.3. Consequences of the fundamental theorem.

(5.3.1) Proposition. Let $\varphi: E_1 \longrightarrow E_2$ be continuous at all, and differentiable at almost all points of the segment S from a to a + h, where $a, h \in E_1$; let E_2^o be separated; $\ell \in L(E_1; E_2)$; B a convex subset of E_2 which is closed in E_2^o. Then we have:

$$\left.\begin{array}{l} \varphi'(x)(h) - \ell(h) \in B \\ \text{for almost all } x \in S \end{array}\right\} \Longrightarrow \varphi(a+h) - \varphi(a) - \ell(h) \in B.$$

Proof. We first consider the special case $\ell = 0$. Let $f: [0,1] \longrightarrow E_2$ be the map defined by

$$f(t) = \varphi(a+th) = (\varphi \circ g)(t), \qquad \text{where}$$

$$g(t) = a + th.$$

$g: [0,1] \longrightarrow E_1$ is differentiable and the chain rule gives

$$f'(t) = f'(t)(1) = (\varphi'(g(t)) \circ g'(t))(1) = \varphi'(g(t))(g'(t)) = \varphi'(g(t))(h).$$

Since $g(t) \in S$, we have by assumption:

$$f^{\bullet}(t) \in B \quad \text{for almost all } t \in [0,1] .$$

Hence the fundamental theorem implies

$$f(1) - f(0) \in B,$$

which is exactly what we want to show, since $\varphi(a+h) - \varphi(a) = f(1)-f(0)$.

The general case $\ell \neq 0$ is reduced to the special case by introducing the map $\psi: E_1 \longrightarrow E_2$ defined by

$$\psi(x) = \varphi(x) - \ell(x). \quad \text{Then}$$

$$\psi'(x)(h) = \varphi'(x)(h) - \ell(h).$$

By applying the result for the special case we have

$$\psi(a+h) - \psi(a) \in B,$$

which is precisely what we wanted to show.

(5.3.2) Remark. Proposition (5.3.1) is also valid under the following weaker hypothesis concerning continuity and differentiability of φ: Assume that $\varphi: E_1^{\#} \longrightarrow E_2^o$ is continuous at all and differentiable at almost all points $x \in S$.

In fact, g being as before, it follows easily (cf (2.6.2)) that $g: [0,1] \longrightarrow E_1^{\#}$ is differentiable and continuous. As before, we apply the fundamental theorem to the map f which is the composite of the two maps

$$[0,1] \xrightarrow{g} E_1^{\#} \xrightarrow{\varphi} E_2^o.$$

Since our approach to calculus is established by means of filters, the appropriate result for later applications is a filter-theoretic version of the fundamental theorem, respectively of proposition (5.3.1.). It will combine in a filter-inequality the main operators of calculus: differentiation, the operator Θ, the taking of closures and of convex hulls. In order to formulate it, we first introduce and discuss some notations.

(5.3.3) Definition. If $A \subset E$ is any subset of a pseudo-topological vector space E, we denote by $\overline{A}$ its closure in E^o (remember that E^o is topological). If $\mathfrak{X}$ is a filter on E, we analogously denote by $\overline{\mathfrak{X}}$ or $(\mathfrak{X})^-$ the filter generated by $\{\overline{X} \mid X \in \mathfrak{X}\}$, which is a filter basis, since $\overline{X_1 \cap X_2} \subset \overline{X}_1 \cap \overline{X}_2$.

Let us denote by $D(E_1;E_2)$ the vector space consisting of the differentiable functions f from E_1 into E_2 which satisfy $f(0) = 0$. The differentiation operator "'" gives a mapping from $D(E_1;E_2)$ into the space of mappings from E_1 into $L(E_1;E_2)$. Therefore we can, for any filter $\mathfrak{F}$ on $D(E_1;E_2)$, consider the filter $\mathfrak{F}'$, which is the image of $\mathfrak{F}$ under this map (or operator) "'". Similarly, $\Theta\mathfrak{F}$ is the image of $\mathfrak{F}$ under the operator Θ introduced in (3.1.1); thus $\Theta\mathfrak{F}$ is a filter on the space of maps from $\mathbb{R} \times E_1$ into E_2.

(5.3.4) Proposition. Let $\mathcal{F}$ be a filter on the space $D(E_1;E_2)$ definited above; $\mathcal{V}$ any filter on $\mathbb{R}$; $\mathfrak{X}$ any filter on E_1; $[0,1]$ the unit interval of $\mathbb{R}$, respectively the filter generated by it; " $\dot{\theta}$ ", " $\circ$ " and " $-$ " the operators defined in (3.1.1), (2.7.2) and (5.3.3). Then

$$\theta \mathcal{F}(\mathcal{V},\mathfrak{X}) \leq (\mathcal{F}'(\mathcal{V}.[0,1].\mathfrak{X})(\mathfrak{X}))^{\circ -}.$$

Proof. Let $\mathfrak{M}$ be any set belonging to the filter on the right hand side of this inequality. Then there exist $V \in \mathcal{V}$, $F \in \mathcal{F}$ and $X \in \mathfrak{X}$ such that the set $B = (F'(V.[0,1].X)(X))^{\circ -}$ is contained in $\mathfrak{M}$:

$\mathfrak{M} \supset (F'(V.[0,1].X)(X))^{\circ -} = B \supset F'(V.[0,1].X)(X).$

Hence

$f'(\vartheta.\lambda.x)(\lambda x) \in \lambda.B$ for $0 \leq \vartheta \leq 1$, $f \in F$, $\lambda \in V, x \in X$.

Since $\lambda.B$ is closed in E° and convex, we can now apply (5.3.1), which yields (since $f(0) = 0$ by assumption):

$f(\lambda.x) \in \lambda.B$ for $f \in F$, $\lambda \in V$, $x \in X$.

Multiplying this by $1/\lambda$ if $\lambda \neq 0$, respectively using that $0 \in B$, we get $\theta f(\lambda,x) \in B$ for $f \in F$, $\lambda \in V$, $x \in X$, or equivalently:

$$B \supset \theta F(V,X).$$

This shows that B (thus a fortiori $\mathfrak{M}$) belongs to the filter $\theta\mathcal{F}(\mathcal{V},\mathfrak{X})$, and the proof is complete.

We shall use (5.3.4) in two special cases, taking as $\mathcal{V}$ either one of

the filters $\mathfrak{W}$ or $[1]$. In the first case we get $\mathfrak{V} . [0,1] . \mathfrak{X} = \mathfrak{W} . \mathfrak{X}$.
In the second case we shall assume that $[0,1] . \mathfrak{X} = \mathfrak{X}$ and we will have:
$\mathfrak{V} . [0,1] . \mathfrak{X} = \mathfrak{X}$ and $\Theta \mathfrak{F}([1], \mathfrak{X}) = \mathfrak{F}(\mathfrak{X})$.

We thus have, $\mathfrak{F}$ being as before, the following two corollaries of (5.3.4):

(5.3.5) Corollary 1.

$$\Theta \mathfrak{F}(\mathfrak{W}, \mathfrak{X}) \leq (\mathfrak{F}'(\mathfrak{W} . \mathfrak{X})(\mathfrak{X}))^{o-}.$$

(5.3.6) Corollary 2. If $\mathfrak{X} = [0,1] . \mathfrak{X}$, then

$$\mathfrak{F}(\mathfrak{X}) \leq (\mathfrak{F}'(\mathfrak{X})(\mathfrak{X}))^{o-}.$$

§ 6. PSEUDO-TOPOLOGIES ON SOME FUNCTION SPACES.

6.1. The spaces $B(E_1;E_2)$, $C_o(E_1;E_2)$ and $L(E_1;E_2)$.

The set $\underline{B}(E_1;E_2)$ of all quasi-bounded maps from E_1 into E_2 (cf.(2.8.4)) is a vector space, since from

$\mathcal{W}.(f_1+f_2)(\mathcal{B}) \leq \mathcal{W}.f_1(\mathcal{B}) + \mathcal{W}.f_2(\mathcal{B})$ (cf. (1.5.2)) and

$\mathcal{W}.(\lambda f)(\mathcal{B}) = \lambda.(\mathcal{W}.f(\mathcal{B}))$ one deduces easily that any linear combination $\lambda_1.f_1 + \lambda_2.f_2$ of two quasi-bounded maps f_1, f_2 is also quasi-bounded.

We denote by $B(E_1;E_2)$ this vector space together with the pseudo-topology caracterized by the following condition:

(6.1.1) $$\mathcal{F}\downarrow B(E_1;E_2) \quad \text{iff} \quad \mathcal{W}.\mathcal{B}\downarrow E_1 \Longrightarrow \mathcal{F}(\mathcal{B})\downarrow E_2.$$

We claim that this definition yields a compatible pseudo-topology on the vector space $\underline{B}(E_1;E_2)$. For this, we have to verify that conditions (2.1.1) are satisfied at the point zero, and that the compatibility conditions (2.4.2) hold. Of the three conditions (2.1.1), only the second one, which demands that

$\mathcal{F}_1 \vee \mathcal{F}_2 \downarrow B(E_1;E_2)$ if $\mathcal{F}_i \downarrow B(E_1;E_2)$ for i = 1,2, is not obvious. But this follows easily, making use of the equality

(6.1.2) $$(\mathcal{F}_1 \vee \mathcal{F}_2)(\mathcal{B}) = \mathcal{F}_1(\mathcal{B}) \vee \mathcal{F}_2(\mathcal{B}),$$

which is a consequence of the set-theoretic equality $(F_1 \cup F_2)(B) = F_1(B) \cup F_2(B)$.

The compatibility conditions (2.4.2) follow respectively from

(a) $(\mathcal{F}_1 + \mathcal{F}_2)(\mathcal{B}) \leq \mathcal{F}_1(\mathcal{B}) + \mathcal{F}_2(\mathcal{B})$ (cf.(1.5.2));

(b) $(\lambda.\mathcal{F})(\mathcal{B}) = \lambda.\mathcal{F}(\mathcal{B})$;

(c) $(\mathcal{V}.\mathcal{F})(\mathcal{B}) = \mathcal{V}.\mathcal{F}(\mathcal{B})$;

(d) $(\mathcal{V}.[f])(\mathcal{B}) = \mathcal{V}.f(\mathcal{B})$.

We remark that for the verification of the fourth compatibility condition it is essential that $\underline{B}(E_1;E_2)$ is not the space consisting of all mappings from E_1 into E_2, but only of the quasi-bounded ones.

We next consider the subspace consisting of the quasi-bounded and equably continuous maps from E_1 into E_2, which was already introduced and denoted by $\underline{C}_o(E_1;E_2)$ in section 2.8. We denote now by $C_o(E_1;E_2)$ the vector space $\underline{C}_o(E_1;E_2)$ together with the pseudo-topology induced by the inclusion. We thus have:

(6.1.3) $$\mathcal{F}\downarrow C_o(E_1;E_2) \quad \text{iff} \quad \mathcal{V}.\mathcal{B}\downarrow E_1 \Longrightarrow \mathcal{F}(\mathcal{B})\downarrow E_2.$$

In the same way, $L(E_1;E_2)$ denotes the pseudo-topological vector space whose underlying space is the space $\underline{L}(E_1;E_2)$ of continuous linear maps from E_1 into E_2, together with the pseudo-topology induced by the inclusion of $\underline{L}(E_1;E_2)$ in $C_o(E_1;E_2)$ or in $B(E_1;E_2)$. Therefore we have:

(6.1.4) $$\mathcal{F}\downarrow L(E_1;E_2) \quad \text{iff} \quad \mathcal{V}.\mathcal{B}\downarrow E_1 \Longrightarrow \mathcal{F}(\mathcal{B})\downarrow E_2 .$$

We shall also need multilinear maps. Let $\underline{L}(E_1,\ldots,E_n;E)$ be the vector space consisting of the equably continuous multilinear maps from $E_1 \times \ldots \times E_n$ into E. We have $\underline{L}(E_1,\ldots,E_n;E) \subset \underline{B}(E_1 \times \ldots \times E_n;E)$, and thus we can consider on $\underline{L}(E_1,\ldots,E_n;E)$ the pseudo-topology induced by that of $B(E_1 \times \ldots \times E_n;E)$. Together with this structure we denote the space of equably continuous multilinear maps by $L(E_1,\ldots,E_n;E)$, and we have therefore:

(6.1.5) $\mathcal{F} \downarrow L(E_1,\ldots,E_n;E)$ iff

$$\forall.\ \mathcal{B}_i \downarrow E_i \quad \text{for } i = 1,\ldots,n \Longrightarrow \mathcal{F}(\mathcal{B}_1,\ldots,\mathcal{B}_n) \downarrow E.$$

The case $E_1 = E_2 = \ldots = E_n$ will be of special interest and we will use the shorter notation

(6.1.6) $$L_n(E_1;E) = L(E_1,\ldots,E_1;E).$$

Besides these structures we can consider the structures which are associated to them by means of the operators " $^{\#}$ " or " $^{\bullet}$ ". It will be convenient not to write the operator at the end, but immediately after the B, C_o, L or L_n; e.g.

(6.1.7) $$C_o^{\#}(E_1;E_2) = (C_o(E_1;E_2))^{\#}.$$

In special cases, the structure of $B(E_1;E_2)$ can be topological:

(6.1.8) Proposition. If E_1 is a normed vector space and E_2 a topological one, then the pseudo-topology of $B(E_1;E_2)$ is the topology of uniform convergence on bounded sets.

Proof. Denoting by $B^*(E_1;E_2)$ the space $\underline{B}(E_1;E_2)$ together with the topology of uniform convergence on bounded sets, we use that this topology is caracterized as follows: $\mathcal{F}\downarrow B^*(E_1;E_2)$ if and only if for each bounded subset B of E_1, $\mathcal{F}(B)\downarrow E_2$. Using lemma (2.5.1), one shows immediately that

$$\mathcal{F}\downarrow B^*(E_1;E_2) \Longleftrightarrow \mathcal{F}\downarrow B(E_1;E_2),$$

which completes the proof.

Combining this result with Proposition (2.8.6) we get now immediately

(6.1.9) Proposition. If E_1 is a finite dimensional vector space with its natural topology and E_2 a normed vector space, then $C_o(E_1;E_2)$ is the space of continuous maps from E_1 into E_2 with the topology of uniform convergence on bounded sets.

(6.1.10) Proposition. If E_1 and E_2 are normed vector spaces, then $L(E_1;E_2)$ is the space of continuous linear maps from E_1 into E_2 together with the topology induced by the usual norm (cf [3]) on $\underline{L}(E_1;E_2)$.

Proof. One combines (6.1.8) with the well known fact that the norm-topology on $\underline{L}(E_1;E_2)$ is the topology of uniform convergence on bounded sets.

(6.1.11) Remark. However, if we assume E_1 and E_2 to be topological, not even $L(E_1;E_2)$ is topological in general. In fact, as it was shown by H.H.Keller (cf. [5]), there does not exist for non-normed vector spaces a topology having the properties which we shall need and which we shall verify in the sequel. The structure used in our theory on $\underline{L}(E_1;E_2)$ is different from the one used by Bastiani [1] and Binz [2] . In the case of locally convex topological vector spaces, our structure is related to a pseudo-topology introduced by H.H.Keller in [5] , using families of seminorms, as follows: the structure of Keller is that of our $L^{\#}(E_1;E_2)$. A detailed discussion will be given in a forthcoming paper.

6.2. Continuity of evaluation maps.

(6.2.1) Proposition. The evaluation map

$$e\colon C_0(E_1;E_2) \times E_1 \longrightarrow E_2,$$

defined by $e(f,x) = f(x)$, is continuous.

Proof. We show continuity at (f,x). So let $\mathfrak{F} \downarrow_f C_0(E_1;E_2)$ and $\mathfrak{X} \downarrow_x E_1$. Then $\mathfrak{F} - f \downarrow C_0(E_1;E_2)$ and $\mathbf{V}.\mathfrak{X} \downarrow E_1$. By (1.5.2) we get

$$\mathfrak{F}(\mathfrak{X}) \leq (\mathfrak{F} - f)(\mathfrak{X}) + f(\mathfrak{X}).$$

The first term on the right side converges to zero on E_2, the second (using that f is continuous) to f(x); hence the sum and a fortiori the left hand side converge to f(x): $\mathfrak{F}(\mathfrak{X}) \downarrow_{f(x)} E_2$, which completes the proof. We remark, that we did not use the equable continuity of f here; but continuity was essential.

(6.2.2) Corollary. The following evaluation maps are continuous:

$e: L(E_1,\dots,E_n;E) \times E_1 \times \dots \times E_n \longrightarrow E;$

$e: L_p(E_1;E) \times E_1 \times \dots \times E_1 \longrightarrow E.$

Since evaluation of multilinear maps is a multilinear mapping, we get by the generalisation of (2.9.2) to multilinear maps:

(6.2.3) Proposition. The evaluation map

$e: L_p{}^{*}(E_1;E) \times E_1{}^{*} \times \dots \times E_1{}^{*} \longrightarrow E^{*}$

is continuous.

The evaluation map of (6.2.1) is not, however, bilinear. But since (2.6.3) implies

$$C_o{}^{*}(E_1;E_2) \times E_1 \leq C_o(E_1;E_2) \times E_1,$$

we get:

(6.2.4) Proposition. The evaluation map

$e: C_o{}^{*}(E_1;E_2) \times E_1 \longrightarrow E_2$

is continuous.

One of the difficulties of the theory is due to the fact that these evaluation maps, such as e.g. the map $e: L(E_1;E_2) \times E_1 \longrightarrow E_2$, fail to be equably continuous in general.

6.3. Continuity of composition maps.

(6.3.1) Proposition. The composition map

$$c: B(E_1;E_2) \times L(E_2;E_3) \longrightarrow B(E_1;E_3),$$

defined by $c(f,\ell) = \ell \circ f$, is continuous.

Proof. Let $\mathfrak{F} \downarrow_g B(E_1;E_2)$ and $\mathfrak{L} \downarrow_k L(E_2;E_3)$. The map c being bilinear, we have, for $f, g \in B(E_1;E_2)$ and $\ell, k \in L(E_2;E_3)$:

$$\ell \circ f - k \circ g = (\ell - k) \circ (f-g) + k \circ (f-g) + (\ell - k) \circ g.$$

This implies, using (1.5.2):

$$c(\mathfrak{F},\mathfrak{L}) - c(g,k) \leq (\mathfrak{L} - k) \circ (\mathfrak{F} - g) + k \circ (\mathfrak{F} - g) + (\mathfrak{L} - k) \circ g.$$

Let now $\mathfrak{W}.\mathfrak{B} \downarrow E_1$. Then, again using (1.5.2):

$$(c(\mathfrak{F},\mathfrak{L}) - c(g,k))(\mathfrak{B}) \leq (\mathfrak{L} - k)(\mathfrak{F} - g)(\mathfrak{B}) + k((\mathfrak{F} - g)(\mathfrak{B})) + (\mathfrak{L} - k)(g(\mathfrak{B})).$$

Since each of the three terms on the right hand side converges to zero on E_3, also the left hand side does. This implies:

$$c(\mathfrak{F},\mathfrak{G}) - c(g,k) \downarrow B(E_1;E_3),$$

which proves the continuity of c at the point (g,k).

Using (2.3.6) we get:

(6.3.2) Corollary. The following composition maps are continuous:

c: $L(E_1,\dots,E_n;E) \times L(E;F) \longrightarrow L(E_1,\dots,E_n;F)$;

c: $L_p(E_1;E_2) \times L(E_2;E_3) \longrightarrow L_p(E_1;E_3)$.

Again, these composition maps fail to be equably continuous in general. However, (6.3.2) implies by (2.9.2) the following

(6.3.3) Proposition. The following composition maps are equably continuous:

c: $L^{*}(E_1,\dots,E_n;E) \times L^{*}(E;F) \longrightarrow L^{*}(E_1,\dots,E_n;F)$;

c: $L_p^{*}(E_1;E_2) \times L^{*}(E_2;E_3) \longrightarrow L_p^{*}(E_1;E_3)$.

6.4. Some canonical isomorphisms.

We first consider the vector spaces $L(\mathbb{R};E)$ and E, which are isomorphic, and we investigate whether they are also homeomorphic.

(6.4.1) Lemma. (a) The canonical isomorphism

$\Phi: L(\mathbb{R};E) \longrightarrow E$,

defined by $\Phi(\ell) = \ell(1)$, is continuous. For its inverse map Ψ, caracterized by $(\Psi(x))(\lambda) = \lambda.x$, we have

(b) $\Psi: E^{*} \longrightarrow L(\mathbb{R};E)$ is continuous;

(c) $\Psi: E \longrightarrow L(\mathbb{R};E)$ is continuous provided that scalar multiplication of E is equably continuous.

Proof. (a) Φ being the evaluation at 1, its continuity follows from (6.2.i).

(b) Let $\mathfrak{X} \downarrow E^{*}$. Hence $\mathfrak{X} \leq \mathfrak{Y} = \mathbb{V}.\mathfrak{Y} \downarrow E$. Thus we get for any $\mathfrak{B}$ with $\mathbb{V}.\mathfrak{B} \downarrow \mathbb{R}$:

$$(\Psi(\mathfrak{X}))(\mathfrak{B}) = \mathfrak{B}.\mathfrak{X} \leq \mathfrak{B}.(\mathbb{V}.\mathfrak{Y}) = (\mathbb{V}.\mathfrak{B}).\mathfrak{Y} \downarrow E,$$

which shows that $\Psi(\mathfrak{X}) \downarrow L(\mathbb{R};E)$. The assertion now follows by (2.8.7).

(c) Similarly we get, if $\mathfrak{X} \downarrow E$ and $\mathbb{V}.\mathfrak{B} \downarrow \mathbb{R}$:

$$(\psi(\mathfrak{X}))(\mathfrak{B}) = \mathfrak{B}.\mathfrak{X},$$

and we deduce from (2.8.8) that this converges to zero on E. The rest goes as before.

From (a) and (b) of the above lemma we get for an arbitrary E, using (2.9.1):

(6.4.2) Proposition. There is a natural (*) linear homeomorphism $L^{*}(\mathbb{R};E) \approx E^{*}$.

From (a) and (c) of lemma (6.4.1) we get:

(6.4.3) Proposition. If scalar multiplication of E is equably continuous, in particular (cf. (2.9.2)) if E is equable, then we have a natural linear homeomorphism:

$$L(\mathbb{R};E) \approx E.$$

(*) We do not discuss the categorical meaning of "natural" or "canonical", so the statement simply means: the isomorphism which we are considering is a homeomorphism.

This in particular implies that $L(\mathbb{R};E)$ is equable if E is equable. We further mention, that from (2.8.8) and (2.5.1) it follows easily, that scalar multiplication on E is equably continuous if and only if $\mathfrak{X} \downarrow E$ implies $I_1 . \mathfrak{X} \downarrow E$; this condition will be satisfied by the so-called admissible spaces considered in § 7 and later.

(6.4.4) Proposition. If E_2 is equable, then we have a natural linear homeomorphism:

$$L(E_1;L(E_2;E_3)) \approx L(E_1,E_2;E_3).$$

Proof. (a) Let $\ell \in L(E_1;L(E_2;E_3))$. We consider the map $\ell \longmapsto \alpha(\ell)$, where $\alpha(\ell) = b$ is the map from $E_1 \times E_2$ into E_3 defined by

(6.4.5) $$b(x_1,x_2) = (\alpha(\ell))(x_1,x_2) = (\ell(x_1))(x_2).$$

This map $b\colon E_1 \times E_2 \longrightarrow E_3$ is obviously bilinear. We show furthermore, that b satisfies the conditions

(6.4.6) $$\mathfrak{X}_1 \downarrow E_1;\ \mathbb{V}.\mathfrak{B}_2 \downarrow E_2 \Longrightarrow b(\mathfrak{X}_1,\mathfrak{B}_2) \downarrow E_3;$$

(6.4.7) $$\mathbb{V}.\mathfrak{B}_1 \downarrow E_1;\ \mathfrak{X}_2 \downarrow E_2 \Longrightarrow b(\mathfrak{B}_1,\mathfrak{X}_2) \downarrow E_3.$$

In fact, (6.4.6) follows easily: since $\ell\colon E_1 \longrightarrow L(E_2;E_3)$ is continuous, $\mathfrak{X}_1 \downarrow E_1$ implies $\ell(\mathfrak{X}_1) \downarrow L(E_2;E_3)$, and hence $(\ell(\mathfrak{X}_1))(\mathfrak{B}_2) \downarrow E_3$. But by (6.4.5), $(\ell(\mathfrak{X}_1))(\mathfrak{B}_2) = b(\mathfrak{X}_1;\mathfrak{B}_2)$. In order to establish (6.4.7), we have to use that E_2 is equable. Therefore $\mathfrak{X}_2 \downarrow E_2$ implies $\mathfrak{X}_2 \leq \mathfrak{Y}_2 = \mathbb{V}.\mathfrak{Y}_2 \downarrow E_2$, and we obtain $b(\mathfrak{B}_1,\mathfrak{X}_2) = (\ell(\mathfrak{B}_1))(\mathfrak{X}_2) \leq (\ell(\mathfrak{B}_1))(\mathbb{V}\mathfrak{Y}_2) = (\ell(\mathbb{V}.\mathfrak{B}_1))(\mathfrak{Y}_2)$.

Here, the filter on the right side converges to zero on E_3, since $\ell(\mathcal{W}.\mathcal{B}_1)\downarrow L(E_2;E_3)$ by the continuity of ℓ and since $\mathcal{W}.\mathcal{Y}_2\downarrow E_2$.

From (6.4.6) and (6.4.7) it follows now by (2.8.8) that $b: E_1\times E_2 \longrightarrow E_3$ is equably continuous, and hence $b\in L(E_1,E_2;E_3)$. We have shown so far that the map $\ell \longmapsto \alpha(\ell) = b$ has its image in $L(E_1,E_2;E_3)$, i.e. we have

(6.4.8) $$\alpha: L(E_1;L(E_2;E_3)) \longrightarrow L(E_1,E_2;E_3).$$

(b) In order to show that α is bijective, we now construct a map

(6.4.9) $$\beta: L(E_1,E_2;E_3) \longrightarrow L(E_1;L(E_2;E_3)).$$

So let $b\in L(E_1,E_2,E_3)$. We first define, for any fixed $x_1\in E_1$, a map $\ell_{x_1} : E_2 \longrightarrow E_3$ by

$$\ell_{x_1}(x_2) = b(x_1,x_2).$$

Since $b: E_1\times E_2 \longrightarrow E_3$ is continuous, $\ell_{x_1} : E_2 \longrightarrow E_3$ is continuous, and we have: $\ell_{x_1} \in L(E_2;E_3)$. We define $\ell = \beta(b) : E_1 \longrightarrow L(E_2;E_3)$ by $\ell(x_1) = \ell_{x_1}$, and we thus have caracterized β (respectively $\beta(b)$) by the equation . . $((\beta(b)(x_1))(x_2) = b(x_1,x_2)$.

$\ell = \beta(b)$ is obviously a linear map from E_1 into $L(E_2;E_3)$.

We want to show that $\ell : E_1 \longrightarrow L(E_2;E_3)$ is continuous. So let $\mathfrak{X}_1\downarrow E_1$. We have to show that $\ell(\mathfrak{X}_1)\downarrow L(E_2;E_3)$, which is equivalent to the condition

$$\mathcal{W}.\mathcal{B}_2\downarrow E_2 \Longrightarrow (\ell(\mathfrak{X}_1))(\mathcal{B}_2)\downarrow E_3,$$

and this holds by (2.8.8), since $\ell(\mathfrak{X}_1)(\mathfrak{B}_2) = b(\mathfrak{X}_1, \mathfrak{B}_2)$ by (6.4.5). We remark that at this point it was essential that $b \in L(E_1,E_2;E_3)$ implies according to the definition of the spaces $L(E_1,E_2;E_3)$ that b is equably continuous.

Having shown that $\ell = \beta(b)$ is linear and continuous, we know now that in fact $\ell = \beta(b) \in L(E_1;L(E_2;E_3))$, and that therefore β is a map as stated in (6.4.9).

It is easy to verify that $\alpha(\beta(b)) = b$ for all $b \in L(E_1,E_2;E_3)$ and $\beta(\alpha(\ell)) = \ell$ for all $\ell \in L(E_1;L(E_2;E_3))$, and thus α is bijective, β being its inverse.

(c) We show that the map α of (6.4.8) is continuous. Since α is linear, we have to verify its continuity at the origin. So let $\mathcal{L} \downarrow L(E_1;L(E_2;E_3))$. In order to show that $\alpha(\mathcal{L}) \downarrow L(E_1,E_2;E_3)$, we have (cf.(6.1.5)) to form $\alpha(\mathcal{L})(\mathfrak{B}_1, \mathfrak{B}_2)$ and to verify that this converges to zero on E_3 provided that $\mathbb{V}.\ \mathfrak{B}_i \downarrow E_i$, i = 1,2. But by (6.4.5) we have $(\alpha(\mathcal{L}))(\mathfrak{B}_1, \mathfrak{B}_2) = (\mathcal{L}(\mathfrak{B}_1))(\mathfrak{B}_2)$, and here $\mathcal{L}(\mathfrak{B}_1) \downarrow L(E_2;E_3)$ and thus $(\mathcal{L}(\mathfrak{B}_1))(\mathfrak{B}_2) \downarrow E_3$.

(d) We finally show that the map $\beta = \alpha^{-1}$ of (6.4.9) is continuous. In fact, let $\mathfrak{F} \downarrow L(E_1,E_2;E_3)$. $\beta(\mathfrak{F}) \downarrow L(E_1;L(E_2;E_3))$ is equivalent to

$$\mathbb{V}.\ \mathfrak{B} \downarrow E_1 \Longrightarrow (\beta(\mathfrak{F}))(\mathfrak{B}_1) \downarrow L(E_2;E_3),$$

and this again is equivalent to

$$\mathbb{V}\ \mathfrak{B}_1 \downarrow E_1,\ \mathbb{V}\ \mathfrak{B}_2 \downarrow E_2 \Longrightarrow ((\beta(\mathfrak{F}))(\mathfrak{B}_1))(\mathfrak{B}_2) \downarrow E_3,$$

and this holds since by the equation caracterizing β one has

$$((\beta(\mathcal{F}))(\mathcal{B}_1))(\mathcal{B}_2) = \mathcal{F}(\mathcal{B}_1, \mathcal{B}_2).$$

This completes the proof of (6.4.4).

(6.4.10) <u>Lemma</u>. $L^{\#}(E_1{}^{\#};E_2) = L^{\#}(E_1{}^{\#};E_2{}^{\#})$.

<u>Proof</u>. (a) We first show that the underlying spaces are the same. If $\ell \in \underline{L}^{\#}(E_1{}^{\#};E_2{}^{\#})$, then $\ell: E_1{}^{\#} \longrightarrow E_2{}^{\#}$ is continuous, and thus a fortiori $\ell: E_1{}^{\#} \longrightarrow E_2$ is continuous, showing that $\ell \in \underline{L}^{\#}(E_1{}^{\#};E_2)$. Conversely, if $\ell \in \underline{L}^{\#}(E_1{}^{\#};E_2)$, we deduce from the continuity of $\ell: E_1{}^{\#} \longrightarrow E_2$ and (2.9.1) the continuity of $\ell: E_1{}^{\#} \longrightarrow E_2{}^{\#}$.

(b) We now show that the structures are the same. One part is immediate: if $\mathcal{L} \downarrow L^{\#}(E_1{}^{\#};E_2{}^{\#})$, then $\mathcal{L} \downarrow L^{\#}(E_1{}^{\#};E_2)$. Suppose conversely that $\mathcal{L} \downarrow L^{\#}(E_1{}^{\#};E_2)$. Then $\mathcal{L} \leq \mathcal{K} = \mathcal{V}.\mathcal{K} \downarrow L(E_1{}^{\#};E_2)$. Let $\mathcal{V}.\mathcal{B}_1 \downarrow E_1{}^{\#}$. Then $\mathcal{K}(\mathcal{B}_1) \downarrow E_2$. But since $\mathcal{K}(\mathcal{B}_1) = (\mathcal{V}.\mathcal{K})(\mathcal{B}_1) = \mathcal{V}.\mathcal{K}(\mathcal{B}_1)$, we have even: $\mathcal{K}(\mathcal{B}_1) \downarrow E_2{}^{\#}$. This shows that $\mathcal{K} \downarrow L(E_1{}^{\#};E_2{}^{\#})$, and since $\mathcal{L} \leq \mathcal{K} = \mathcal{V}.\mathcal{K}$ it follows that $\mathcal{L} \downarrow L^{\#}(E_1{}^{\#};E_2{}^{\#})$.

(6.4.11) <u>Proposition</u>. If E_1 and E_2 are equable, then we have a natural linear homeomorphism:

$$L^{\#}(E_1;L^{\#}(E_2;E_3)) \approx L^{\#}(E_1,E_2;E_3).$$

Proof. The map α of (6.4.8) being linear and continuous, we deduce by (2.9.1) that also

$$\alpha: L^{\#}(E_1;L(E_2;E_3)) \longrightarrow L^{\#}(E_1,E_2;E_3)$$

is linear. The same holds for the map $\beta = \alpha^{-1}$. Now using that $E = E_1^{\#}$ by assumption, the result follows, since by lemma (6.4.10) we have $L^{\#}(E_1;L(E_2;E_3)) = L^{\#}(E_1;L^{\#}(E_2;E_3))$.

(6.4.12) Proposition. If E_1 is equable, then there are natural linear homeomorphisms as follows:

$$L_p(E_1;L_q(E_1;E_2)) \approx L_{p+q}(E_1;E_2);$$
$$L_p^{\#}(E_1;L_q^{\#}(E_1;E_2)) \approx L_{p+q}^{\#}(E_1;E_2).$$

The proofs are completely analogous to those of propositions (6.4.4) and (6.4.11). Instead of (2.8.8) one has to use the analogous result which states that a multilinear map $\ell: E_1 \times \ldots \times E_n \longrightarrow E$ is equably continuous if and only if $\ell(\mathfrak{X}_1,\ldots,\mathfrak{X}_n) \downarrow E$ provided that one of the filters $\mathfrak{X}_1,\ldots,\mathfrak{X}_n$ converges to zero, the others being quasi-bounded. And at the place of lemma (6.4.10) one has to establish the corresponding equality

$$L_p^{\#}(E_1^{\#};E_2) = L_p^{\#}(E_1^{\#};E_2^{\#}).$$

The map which we call natural is of course the following: to $\ell \in L_p(E_1;L_q(E_1;E_2))$ we associate $\alpha(\ell) \in L_{p+q}(E_1;E_2)$, $\alpha(\ell)$ being caracterized by

$$(\alpha\ell)(x_1,\ldots,x_{p+q}) = (\ell(x_1,\ldots,x_p))(x_{p+1},\ldots,x_{p+q}).$$

(6.4.13) <u>Proposition</u>. The map $(f_1,f_2) \longmapsto [f_1,f_2]$ (cf.(1.3.2)) yields natural linear homeomorphisms as follows:

$$C_o(E;E_1) \times C_o(E;E_2) \approx C_o(E;E_1 \times E_2);$$

$$L_p(E;E_1) \times L_p(E;E_2) \approx L_p(E;E_1 \times E_2).$$

<u>Proof</u>.(a) We first show that $(f_1,f_2) \in C_o(E;E_1) \times C_o(E;E_2)$ implies $[f_1,f_2] \in C_o(E;E_1 \times E_2)$. By (1.3.3) and (1.5.1) we have

$$[f_1,f_2]\ (\mathfrak{B}) = (f_1 \times f_2)(\,d(\mathfrak{B})) \leq (f_1 \times f_2)(\mathfrak{B} \times \mathfrak{B}) = f_1(\mathfrak{B}) \times f_2(\mathfrak{B}).$$

Thus, if $\mathcal{V}.\mathfrak{B} \downarrow E$, the filter on the right hand side and hence a fortiori $[f_1,f_2]\ (\mathfrak{B})$ is quasi-bounded, which shows that $[f_1,f_2]$ is quasi-bounded. From the equality

$$\Delta[f_1,f_2]\ (a,x) = (\Delta f_1(a,x),\ \Delta f_2(a,x))$$

we deduce similarly that

$$\Delta[f_1,f_2]\ (\mathfrak{A},\mathfrak{X}) \leq \Delta f_1(\mathfrak{A},\mathfrak{X}) \times \Delta f_2(\mathfrak{A},\mathfrak{X}),$$

and from this it is obvious that the equable continuity of f_1 and f_2 implies that of $[f_1,f_2]$. We thus have shown that $(f_1,f_2) \longmapsto [f_1,f_2]$ induces a map

(6.4.14) $$\alpha\colon C_o(E;E_1) \times C_o(E;E_2) \longrightarrow C_o(E;E_1 \times E_2).$$

(b) Let, conversely, $f \in C_o(E;E_1 \times E_2)$. Then we conclude by (2.8.7) and (2.8.5) that $\pi_i \circ f$, where $\pi_i : E_1 \times E_2 \longrightarrow E_i$ is the projection, lies in $C_o(E;E_i)$, $i = 1,2$. Thus $f \longmapsto (\pi_1 \circ f,\ \pi_2 \circ f)$ yields a map

(6.4.15) $$\beta\colon C_o(E;E_1 \times E_2) \longrightarrow C_o(E;E_1) \times C_o(E;E_2),$$

and β is obviously the inverse of α. Both maps α, β are linear, and so it only remains to show that they are continuous, or equivalently: continuous at the origin.

(c) Let $\mathfrak{F} \downarrow C_o(E;E_1) \times C_o(E;E_2)$. Then $\mathfrak{F} \leq \mathfrak{F}_1 \times \mathfrak{F}_2$, where $\mathfrak{F}_i \downarrow C_o(E;E_i)$. If $\mathcal{W}.\mathcal{B} \downarrow E$, we have therefore $\mathfrak{F}_i(\mathcal{B}) \downarrow E_i$, and further, since $(\alpha\mathfrak{F})(\mathcal{B}) \leq \mathfrak{F}_1(\mathcal{B}) \times \mathfrak{F}_2(\mathcal{B})$, also $(\alpha\mathfrak{F})(\mathcal{B}) \downarrow E_1 \times E_2$. This shows that $\alpha\mathfrak{F} \downarrow C_o(E;E_1 \times E_2)$ and hence the map (6.4.14) is continuous.

(d) Let $\mathcal{G} \downarrow C_o(E;E_1 \times E_2)$. We have to show that $\beta\mathcal{G} \downarrow C_o(E;E_1) \times C_o(E;E_2)$, or equivalently: $\omega_i(\beta\mathcal{G}) \downarrow C_o(E;E_i)$, where ω_i, $i = 1,2$, are the projection maps of the product $C_o(E;E_1) \times C_o(E;E_2)$. Since $\omega_i(g) = \pi_i \circ g$ we get

$$\omega_i(\beta\mathcal{G})(\mathcal{B}) = \pi_i(\mathcal{G}(\mathcal{B})) \downarrow E_i,$$

which ends the proof of the first homeomorphism. The second result, concerning p-linear maps, is a corollary; one only has to remark that $[f_1, f_2]$ is p-linear if and only if f_1 and f_2 are p-linear and that $L_p(E;E_i)$ has the structure of subspace of $C_o(E \times \ldots \times E;E_i)$.

(6.4.16) Proposition. If $\ell : E_1 \longrightarrow E_2$ is linear and continuous, then $\ell_*: C_o(E;E_1) \longrightarrow C_o(E;E_2)$ defined by $\ell_*(f) = \ell \circ f$ also is linear and continuous.

By (2.8.5) and (2.8.7) $\ell \circ f \in C_o(E;E_2)$. ℓ_* is obviously linear. Let now $\mathfrak{F} \downarrow C_o(E;E_1)$, $\forall \mathfrak{B} \downarrow E$. Then $\mathfrak{F}(\mathfrak{B}) \downarrow E_1$ and $\ell(\mathfrak{F}(\mathfrak{B})) = (\ell \circ \mathfrak{F})(\mathfrak{B}) \downarrow E_2$, hence $\ell \circ \mathfrak{F} = \ell_*(\mathfrak{F}) \downarrow C_o(E;E_2)$.

(6.4.17) <u>Corollary</u>. If $\ell : E_1 \longrightarrow E_2$ is a linear homeomorphism, then so is $\ell_* : C_o(E;E_1) \longrightarrow C_o(E;E_2)$.

§ 7. THE CLASS OF ADMISSIBLE VECTOR SPACES.

In order to obtain many of the deeper results of calculus in pseudo-topological vector spaces, some restrictions concerning the spaces are necessary. We discuss these conditions and call "admissible" those pseudo-topological vector spaces which satisfy them. Since calculus makes use of various constructions yielding new spaces from given ones (such as e.g. $L(E_1;E_2)$ or $C_o^{\#}(E_1;E_2)$ from E_1, E_2), it will be important to know whether these constructions yield admissible spaces if applied to admissible ones. We shall verify that this is the case.

7.1. The admissibility conditions.

We recall (cf.(5.3.3) and (2.7.1)) that for any filter $\mathfrak{X}$ on a pseudo-topological vector space E, we denoted by $\overline{\mathfrak{X}}$ or $(\mathfrak{X})^{-}$ the filter generated by the closures (in E^o) of the sets of $\mathfrak{X}$, and by $\hat{\mathfrak{X}}$ or $(\mathfrak{X})^{\wedge}$ the filter generated by the convex hulls of the sets of $\mathfrak{X}$.

(7.1.1) <u>Definition</u>. A pseudo-topological vector space E is called <u>admissible</u> iff it satisfies the following three conditions:

(a) E^0 is separated;

(b) $\mathfrak{X} \downarrow E \Longrightarrow \overline{\mathfrak{X}} \downarrow E$;

(c) $\mathfrak{X} \downarrow E \Longrightarrow \hat{\mathfrak{X}} \downarrow E$.

From (2.7.2) and (2.7.3) it follows at once that we have the following

(7.1.2) Lemma. Condition (c) is equivalent to the condition

(c') $\mathfrak{X} \downarrow E \Longrightarrow \mathfrak{X}^0 \downarrow E$.

We further remark that condition (a) implies that E is separated, because by (2.7.10) $\mathfrak{X} \downarrow_x E$ implies $\mathfrak{X} \downarrow_x E^0$. Hence the admissible spaces satisfy in particular the condition made at the beginning of 3.2.

Not even topological vector spaces always satisfy condition (b). To see this, choose a separated topological vector space E for which $\hat{\mathcal{U}} = [E]$, $\mathcal{U}$ denoting the neighborhood filter of zero in E (cf. e.g. [8], p. 161). Then $[0] \downarrow E$; but $\overline{[0]} \downarrow E$ is not true (note that closures have to be taken, according to the definition of $\overline{\mathfrak{X}}$, with respect to E^0, so that $\overline{0} = E$).

(7.1.3) Proposition. Each locally-convex separated topological vector space E is admissible.

<u>Proof</u>. We first remark that for topological vector spaces, condition (c) is equivalent to the classical condition of local convexity. Hence E satisfies (c); and since (cf.(2.7.10)) $E = E^{0}$, also (a) is satisfied. Finally (b) holds, since for $E = E^{0}$ it is well known that the neighborhood filter $\mathfrak{U}$ of zero in E satisfies $\mathfrak{U} = \overline{\mathfrak{U}}$.

(7.1.4) <u>Corollary</u>. If E is admissible, then also E^{0} is admissible.

<u>7.2. Admissibility of $E^{\sharp}$.</u>

(7.2.1) <u>Lemma</u>. If $\mathbb{W}.\mathfrak{X} = \mathfrak{X}$, then $\mathbb{W}.\overline{\mathfrak{X}} = \overline{\mathfrak{X}}$.

<u>Proof</u>. Let $\mathfrak{M} \in \overline{\mathfrak{X}} = \overline{\mathbb{W}.\mathfrak{X}}$. Then $\mathfrak{M} \supset \overline{V.X}$, where $V \in \mathbb{W}$ and $X \in \mathfrak{X}$. Since scalar multiplication of E^{0} is continuous, we get

$$\mathfrak{M} \supset \overline{V.X} \supset \overline{V} . \overline{X} \in \overline{\mathbb{W}} . \overline{\mathfrak{X}} ,$$

which shows, since $\overline{\mathbb{W}} = \mathbb{W}$, that $\mathfrak{M} \in \mathbb{W}. \overline{\mathfrak{X}}$. Suppose, conversely, that $\mathfrak{M} \in \mathbb{W}. \overline{\mathfrak{X}}$. Then $\mathfrak{M} \supset I_{\delta}.\overline{X}$, where by (2.5.2) X can be chosen such that $X = I_{1}.X$. Since multiplication by δ is a homeomorphism of E^{0}, we get :

$$\mathfrak{M} \supset I_{\delta}.\overline{X} \supset \delta.\overline{I_{1}.X} = \overline{\delta.I_{1}.X} = \overline{I_{\delta}.X} \in \overline{\mathbb{W}.\mathfrak{X}} = \overline{\mathfrak{X}} ,$$

which shows that $\mathfrak{M} \in \overline{\mathfrak{X}}$.

(7.2.2) <u>Lemma</u>. If $\mathbb{W}.\mathfrak{X} = \mathfrak{X}$, then $\mathbb{W}.\hat{\mathfrak{X}} = \hat{\mathfrak{X}}$.

Proof. Using standard arguments concerning convexity, one first shows that $I_1.X = X$ implies $I_1.\hat{X} = \hat{X}$. From this, the result follows easily, again making use of (2.5.2).

(7.2.3) Proposition. If E is admissible, then $E^{\#}$ is also admissible.

Proof. We verify that each of the three admissibility conditions carries over from E to $E^{\#}$.

(a) From (2.6.3) we get $(E^{\#})^0 \leq E^0$. Hence E^0 separated implies $(E^{\#})^0$ separated.

(b) For $A \subset E$, we denote by $\bar{A}$ the closure of A with respect to E^0 and by $\bar{A}^{\#}$ the closure with respect to $(E^{\#})^0$. Correspondingly $\bar{\mathfrak{X}}$ and $\bar{\mathfrak{X}}^{\#}$ for a filter $\mathfrak{X}$ on E. Let $\mathfrak{X} \downarrow E^{\#}$. Hence there is $\mathfrak{Y}$ with $\mathfrak{X} \leq \mathfrak{Y} = \mathfrak{W}\mathfrak{Y} \downarrow E$. Using (7.2.1) and the admissibility of E, we get: $\bar{\mathfrak{X}} \leq \bar{\mathfrak{Y}} = \mathfrak{W}\bar{\mathfrak{Y}} \downarrow E$. This shows that

(7.2.4) $$\bar{\mathfrak{X}} \downarrow E^{\#}$$

and a fortiori that $\bar{\mathfrak{X}}^{\#} \downarrow E^{\#}$, because $\bar{\mathfrak{X}}^{\#} \leq \bar{\mathfrak{X}}$.

(c) Let $\mathfrak{X} \downarrow E^{\#}$. As before: $\mathfrak{X} \leq \mathfrak{Y} = \mathfrak{W}.\mathfrak{Y} \downarrow E$, and by (7.2.2) we get: $\hat{\mathfrak{X}} \leq \hat{\mathfrak{Y}} = \mathfrak{W}.\hat{\mathfrak{Y}}$, where $\hat{\mathfrak{Y}} \downarrow E$ since E is admissible. Therefore we have $\hat{\mathfrak{X}} \downarrow E^{\#}$.

7.3. Admissibility of subspaces, direct products and projective limits.

(7.3.1) Lemma. Supose that the spaces E_i, $i \in I$, are admissible; that the maps

$$f_i : E \longrightarrow E_i \ , \ i \in I$$

are linear and such that for each $x \neq 0$, $x \in E$, there exists $k \in I$ with $f_k(x) \neq 0$; and that E has the structure induced by these maps. Then also E is admissible.

Proof. We show that each of the admissibility conditions carries over from the spaces E_i to E.

(a) The maps $f_i : E \longrightarrow E_i$ being continuous and linear, we know by (2.9.1) that the maps

$$f_i : E^o \longrightarrow E_i^o$$

are also continuous. Let $x \in E$, $x \neq 0$., Choose $k \in I$ such that $f_k(x) = x_k \neq 0$. E_k^o being separated, we can in E_k^o choose a neighborhood U_k of x_k with $0 \notin U_k$. Then $U = f_k^{-1}(U_k)$ is a neighborhood of x in E^o, and $0 \notin U$. This shows that E^o is separated.

(b) Let $\mathfrak{X} \downarrow E$. Then $f_i(\mathfrak{X}) \downarrow E_i$ for all $i \in I$, and hence we have

$$\overline{f_i(\mathfrak{X})} \downarrow E_i \qquad \text{for all } i \in I.$$

But the continuity of $f_i : E^o \longrightarrow E_i^o$ implies that $f_i(\overline{X}) \subset \overline{f_i(X)}$ and therefore $f_i(\overline{\mathfrak{X}}) \leq \overline{f_i(\mathfrak{X})}$, closures being taken in E^o resp. E_i^o. We thus have a fortiori:

$$f_i(\overline{\mathfrak{X}}) \downarrow E_i \qquad \text{for all } i \in I,$$

which yields (cf.(2.3.2)) $\overline{\mathfrak{X}} \downarrow E$.

(c) Let $\mathfrak{X} \downarrow E$. Then we have $f_i(\mathfrak{X}) \downarrow E_i$ and

$$(f_i(\mathfrak{X}))^{\wedge} \downarrow E_i \qquad \text{for all } i \in I.$$

Now the linearity of $f_i : E \longrightarrow E_i$ implies that $f_i(\hat{X}) \subset (f_i(X))^{\wedge}$ and therefore $f_i(\hat{\mathfrak{X}}) \leq (f_i(\mathfrak{X}))^{\wedge}$, and we have a fortiori

$$f_i(\hat{\mathfrak{X}}) \downarrow E_i \qquad \text{for all } i \in I,$$

showing that $\hat{\mathfrak{X}} \downarrow E$.

(7.3.2) <u>Proposition</u>. Subspaces, direct products and projective limits of admissible spaces are also admissible.

<u>Proof</u>. One easily checks that in each of these cases, lemma (7.3.1) can be applied.

<u>7.4. Admissibility of $B(E_1;E_2)$, $C_o(E_1;E_2)$, $L_p(E_1;E_2)$.</u>

(7.4.1) <u>Proposition</u>. If E_2 is admissible, then $B(E_1;E_2)$ is also admissible.

<u>Proof</u>. Again we show separately that each of the admissibility conditions carries over from E_2 to $B(E_1;E_2)$.

(a) Let $f_o \in B(E_1;E_2)$, $f \neq 0$. Then we can choose $x_o \in E_1$ with $y_o = f_o(x_o) \neq 0$. In E_2^o there exists a neighborhood U of 0 with $y_o \notin U$. Let φ be the evaluation at x_o, i.e. the map defined by

$$\varphi(f) = f(x_o).$$

If $\mathfrak{F} \downarrow B(E_1;E_2)$, then $\varphi(\mathfrak{F}) = \mathfrak{F}(x_o) \downarrow E_2$ since $\mathfrak{V}.x_o \downarrow E_1$. φ being linear, we conclude by (2.8.7) that $\varphi : B(E_1;E_2) \longrightarrow E_2$ is continuous, and hence (cf.(2.9.1))

(7.4.2) $$\varphi : B^o(E_1;E_2) \longrightarrow E_2^o$$

is also continuous (*). Therefore $V = \varphi^{-1}(U)$ is in $B^o(E_1;E_2)$ a neighborhood of 0. Furthermore, $f_o \notin V$, because $f_o \in V$ would yield $y_o = f_o(x_o) = \varphi(f_o) \in \varphi(V) = \varphi(\varphi^{-1}(U)) \subset U$, which contradicts $y_o \notin U$. This proves that $B^o(E_1;E_2)$ is separated.

(b) We first establish, for any filters $\mathfrak{X}$ resp. $\mathfrak{F}$ on E_1 resp. $B(E_1;E_2)$, the following relation:

(7.4.3) $$\overline{\mathfrak{F}(\mathfrak{X})} \geq \overline{\mathfrak{F}}(\mathfrak{X}) ,$$

which follows if we show that for subsets X resp. F of E_1 resp. $B(E_1;E_2)$ one has

(7.4.4) $$\overline{F(X)} \supset \overline{F}(X).$$

So let $y \in \overline{F}(X)$. Hence $y = f_o(x_o)$, where $f_o \in \overline{F}$ and $x_o \in X$. Using again the continuity of the map (7.4.2), we have $\varphi(\overline{F}) \subset \overline{\varphi(F)}$. We thus obtain $y = f_o(x_o) = \varphi(f_o) \in \varphi(\overline{F}) \subset \overline{\varphi(F)} = \overline{F(x_o)} \subset \overline{F(X)}$, which establishes (7.4.4).

Assume now that $\mathfrak{F} \downarrow B(E_1;E_2)$. Then we have, for any quasi-bounded filter $\mathfrak{B}$ on E_1, $\mathfrak{F}(\mathfrak{B}) \downarrow E_2$. Since E_2 is admissible, we conclude that $\overline{\mathfrak{F}(\mathfrak{B})} \downarrow E_2$ and hence by (7.4.3) also $\overline{\mathfrak{F}}(\mathfrak{B}) \downarrow E_2$. But this shows that $\overline{\mathfrak{F}} \downarrow B(E_1;E_2)$.

(*) We remark that the evaluation e: $B(E_1;E_2) \times E_1 \longrightarrow E_2$ is not continuous in general (cf.(6.2.1)).

(c) $\mathfrak{F}$, $\mathfrak{X}$, F and X being as before, we first remark that

$$\hat{F}(X) \subset (F(X))^{\wedge},$$

a relation concerning convex hulls which is easily checked.

From that we obtain

(7.4.5) $$\hat{\mathfrak{F}}(\mathfrak{X}) \leq (\mathfrak{F}(\mathfrak{X}))^{\wedge}.$$

Let now $\mathfrak{F} \downarrow B(E_1;E_2)$. Then we have, for any quasi-bounded filter $\mathfrak{B}$ on E_1, $\mathfrak{F}(\mathfrak{B}) \downarrow E_2$, from which we get $(\mathfrak{F}(\mathfrak{B}))^{\wedge} \downarrow E_2$ and thus by (7.4.5) a fortiori $\hat{\mathfrak{F}}(\mathfrak{B}) \downarrow E_2$. This shows that $\hat{\mathfrak{F}} \downarrow B(E_1;E_2)$, and the proof of proposition (7.4.1) is complete.

(7.4.6) Proposition. If E_2 is admissible, then also the spaces $C_o(E_1;E_2)$, $C_o^{*}(E_1;E_2)$, $L_p(E_1;E_2)$, $L_p^{*}(E_1;E_2)$ are admissible.

Proof. $C_o(E_1;E_2)$ resp. $L_p(E_1;E_2)$ are subspaces of $B(E_1;E_2)$ resp. $B(E_1 x \ldots x E_1;E_2)$; hence their admissibility results from (7.4.1) and (7.3.2). The rest follows from (7.2.3).

§ 8. PARTIAL DERIVATIVES AND DIFFERENTIABILITY.

8.1. Partial derivatives.

Let $f : E_1 \times E_2 \longrightarrow E_3$. For any fixed $(a_1, a_2) \in E_1 \times E_2$, we can consider the partial mappings $x_1 \longmapsto f(x_1, a_2)$ and $x_2 \longmapsto f(a_1, x_2)$. If these mappings are differentiable at a_1 resp. a_2, f is called partially differentiable at (a_1, a_2), the derivatives being denoted by $D_1 f(a_1, a_2)$ resp. $D_2 f(a_1, a_2)$.

(8.1.1) *Proposition.* If $f: E_1 \times E_2 \longrightarrow E_3$ is differentiable at (a_1, a_2), then f is partially differentiable at (a_1, a_2),

(8.1.2) $Df(a_1, a_2) \cdot (t_1, t_2) = D_1 f(a_1, a_2) . t_1 + D_2 f(a_1, a_2) . t_2$;

and continuity of Df implies continuity of $D_1 f$ and $D_2 f$.

Proof. $x_1 \longmapsto f(x_1, a_2)$ is the composite of the two mappings $x_1 \longmapsto (x_1, a_2)$ and $(x_1, a_2) \longmapsto f(x_1, a_2)$. The first is differentiable by (4.2.1), (3.2.3) and (4.4.5), it's derivative at any point being $i_1 : t_1 \longmapsto (t_1, 0)$. Hence we get by (3.3.1):

(8.1.3) $D_1 f(a_1, a_2) = Df(a_1, a_2) \circ i_1$. Analogously

$D_2 f(a_1, a_2) = Df(a_1, a_2) \circ i_2$, where

$i_2(t_2) = (0, t_2)$.

Since $(t_1, t_2) = i_1(t_1) + i_2(t_2)$, we get (8.1.2). The continuity assertion for $D_1 f$ follows from (8.1.3), using that by (6.3.2) the mapping $c : L(E_1; E_1 \times E_2) \times L(E_1 \times E_2; E_3) \longrightarrow L(E_1; E_3)$ is continuous. Similarly for $D_2 f$.

8.2. A sufficient condition for (total) differentiability.

In the preceding paragraph the definition of admissible vector spaces was given. They now enter essentially in the proof of the following theorem whose classical version is well known.

(8.2.1) Theorem. Let U be an $E_1 \times E_2$-neighborhood of $(0,0) \in E_1 \times E_2$ and E_3 an admissible vector space. Suppose that $f: E_1 \times E_2 \longrightarrow E_3$ is partially differentiable in $(a_1,a_2) + U$ and that $D_2 f: E_1 \times E_2 \longrightarrow L(E_2;E_3)$ is continuous at (a_1,a_2) (*). Then f is differentiable at (a_1,a_2) and

$Df(a_1,a_2) \cdot (t_1,t_2) = D_1 f(a_1,a_2) \cdot t_1 + D_2 f(a_1,a_2) \cdot t_2$.

Further continuity of $D_1 f$ and $D_2 f$ implies continuity of Df.

Proof. $f(a_1+x_1, a_2+x_2) - f(a_1,a_2) =$

(8.2.2) $D_1 f(a_1,a_2) \cdot x_1 + D_2 f(a_1,a_2) \cdot x_2 + (r_2 + r_1 \circ \pi_1) \cdot (x_1,x_2)$,

where $r_1(x_1) = f(a_1+x_1,a_2) - f(a_1,a_2) - D_1 f(a_1,a_2) \cdot x_1$ and

$r_2(x_1,x_2) = f(a_1+x_1,a_2+x_2) - f(a_1+x_1,a_2) - D_2 f(a_1,a_2) \cdot x_2$.

Since by assumption $r_1 \in R(E_1;E_3)$, we get $r_1 \circ \pi_1 \in R(E_1 \times E_2;E_3)$ by (3.1.6).

(*) (1) For $(x_1,x_2) \notin (a_1,a_2) + U$, one can define $D_2 f$ arbitrarily (see also 3.4.1).

(2) E and $E^{\#}$ have the same quasi bounded filters. Hence the following proof also shows that the theorem is true under the weaker hypothesis: U is an $E_1^{\#} \times E_2^{\#}$-neighborhood and $D_2 f: E_1^{\#} \times E_2^{\#} \longrightarrow L(E_2;E_3)$ is continuous.

In order to show that $r_2 \in R(E_1 \times E_2; E_3)$, let $\mathbb{V}\mathcal{B} \downarrow E_1 \times E_2$. Then $\mathcal{B} \leq \mathcal{B}_1 \times \mathcal{B}_2$, where (for i = 1,2) $\mathcal{B}_i = I.\,\pi_i(\mathcal{B})$, I the interval $[0,1] \subset \mathbb{R}$. But $\mathbb{V}\mathcal{B}_i = \mathbb{V}.I.\,\pi_i(\mathcal{B}) = \pi_i(\mathbb{V}.\mathcal{B}) \downarrow E_i$ by continuity of the projections, hence by the continuity of $D_2 f$ at (a_1, a_2):

$D_2 f(a_1 + \mathbb{V}\mathcal{B}_1, a_2 + \mathbb{V}\mathcal{B}_2) - D_2(a_1, a_2) \downarrow L(E_2; E_3)$.

Let $\varphi_{x_1}(x_2) = r_2(x_1, x_2)$. Then for $(x_1, x_2) \in U$, $\varphi_{x_1}'(x_2) =$

$D_2 f\ (a_1 + x_1, a_2 + x_2) - D_2 f(a_1, a_2)$.

Further $\varphi_{x_1}(0) = 0$, hence by (5.3.5)

$$(\varphi_{\mathbb{V}\mathcal{B}_1}'(\mathbb{V}\mathcal{B}_2)\ .\ \mathcal{B}_2)^{0-} \geq \Theta\ \varphi_{\mathbb{V}\mathcal{B}_1}\ .\ (\mathbb{V}, \mathcal{B}_2).$$

Observe now that

$$(\varphi_{\mathbb{V}\mathcal{B}_1}'(\mathbb{V}\mathcal{B}_2)\ .\ \mathcal{B}_2)^{0-} = ((D_2 f(a_1 + \mathbb{V}\mathcal{B}_1, a_2 + \mathbb{V}\mathcal{B}_2) - D_2 f(a_1, a_2)).\ \mathcal{B}_2)^{0-} \downarrow E_3$$

by the admissibility assumption of E_3 and the continuity of $D_2 f$ at (a_1, a_2).

Further

$$\Theta\ \varphi_{\mathbb{V}\mathcal{B}_1}.(\mathbb{V}, \mathcal{B}_2) \geq \Theta\ r_2(\mathbb{V}, \mathcal{B}_1 \times \mathcal{B}_2) \geq .\Theta\ r_2(\mathbb{V}, \mathcal{B})$$

which proves that $r_2 \in R(E_1 \times E_2; E_3)$. Hence by (3.1.4): $r_2 + r_1 \circ \pi_1 \in R(E_1 \times E_2; E_3)$, and (8.2.2) shows that f is differentiable at (a_1, a_2). The continuity assertion for Df follows from (6.3.2), because

$$Df(x_1, x_2) = D_1 f(x_1, x_2) \circ \pi_1 + D_2 f(x_1, x_2) \circ \pi_2.$$

§ 9. HIGHER DERIVATIVES.

9.1. f" and the symmetry of f"(x).

If $u : E_1 \times E_2 \longrightarrow E_3$ is bilinear and continuous, we already know that u is differentiable at every point (4.2.3), but we are not able to prove that $Du : E_1 \times E_2 \longrightarrow L(E_1 \times E_2; E_3)$ is continuous unless E_1 and E_2 are equable vector spaces.

(9.1.1) From now on we suppose that $E, E_1, E_2, \ldots$ are always equable and admissible vector spaces.

(9.1.2) Definition. A map $f: E_1 \longrightarrow E_2$ of equable and admissible vector spaces is called twice differentiable at a point a, iff $Df: E_1 \longrightarrow L^{\#}(E_1;E_2)$ exists in a E-neighborhood of a (cf. 3.4) and is differentiable at a.

We write $D^2f(a)$ instead of $D(Df)(a)$ and remark that thus $D^2f(a) \in L^{\#}(E_1;L^{\#}(E_1;E_2))$. The element which corresponds to $D^2f(a)$ in the natural isomorphism (6.4.2) is denoted by $f''(a)$, and we shall write $f''(a).(s,t)$ instead of $(f''(a))(s,t)$. $f''(a)$ is a bilinear map : $f''(a) \in L_2^{\#}(E_1;E_2)$.

(9.1.3) Proposition. If $f: E_1 \longrightarrow E_2$ is twice differentiable at a, then $f''(a)$, i.e. the bilinear mapping $(s,t) \longmapsto f''(a).(s,t)$, is symmetric (*).

* The same is true under the weaker condition that $f: E_1 \longrightarrow E_2^{o}$ is twice differentiable.

Proof. Let $(s,t) \in E_1 \times E_2$, $\lambda \in \mathbb{R}$ and consider the mapping

$g : [0,1] \longrightarrow E_2$ defined by $g(\xi) = f(a + \lambda\xi . s + \lambda t) - f(a + \lambda\xi . s)$.

Let $I_1 = [0,1]$. By assumption Df exists in a E_1-neighborhood U of a. Hence $a + \mathbb{V}(I_1 . s + t) \ni U$, because $\mathbb{V}[I_1 . s + t] \leq \mathbb{V}s + \mathbb{V}t \downarrow E_1$.

In other words, there exists $\varepsilon_1 > 0$ such that

$a + \lambda\xi . s + \lambda . t \in U$ for $|\lambda| \leq \varepsilon_1$ and $\xi \in [0,1]$, and therefore g is differentiable in $[0,1]$ for each $|\lambda| \leq \varepsilon_1$.

By (3.3.1) and (4.3.2), we get

$g'(\xi) = (Df(a + \lambda\xi s + \lambda t) - Df(a + \lambda\xi s)) . \lambda s$.

Inside any neighborhood of zero in E_2^0 we can choose a neighborhood W which is absolutely convex and closed; this means that

$$-W = W = \hat{W} = \overline{W} .$$

We show that there is $\varepsilon_2 > 0$ with $\varepsilon_2 \leq \varepsilon_1$ such that for $|\lambda| \leq \varepsilon_2$:

(9.1.4) $\quad (Df(a + \lambda\xi s + \lambda t) - Df(a) - D^2 f(a) . (\lambda\xi s + \lambda t)) . s \in \lambda W$

and $(Df(a + \lambda\xi s) - Df(a) - D^2 f(a) . (\lambda\xi s)) . s \in \lambda W$.

From the differentiability of Df at a we conclude also that

$\Theta RDf(a) . (\mathbb{V}, [I_1 . s + t]) \downarrow L(E_1 ; E_2)$ hence, because $\mathbb{V}s \downarrow E_1$,

$\Theta RDf(a) . (\mathbb{V}, [I_1 . s + t]) . s \leq [W]$ which yields (9.1.4).

Now we get by multiplying with λ and subtracting

$g'(\xi) . 1 - ((D^2 f(a) . \lambda t) \circ u_{\lambda s}) . 1 \in 2\lambda^2 . W$ where $u_{\lambda s} : \mathbb{R} \longrightarrow E_2^0$ is defined by $u_{\lambda s}(\eta) = \eta . \lambda s$.

By (5.3.1) we get

$g(1) - g(0) - (D^2f(a).\lambda t).\lambda s \in 2\lambda^2.W$.

$g(1) - g(0)$ being symmetric in s,t, we can find $\varepsilon_3 > 0$ with $\varepsilon_3 \leq \varepsilon_2$, such that for each $|\lambda| \leq \varepsilon_3$:

$g(1) - g(0) - (D^2f(a).\lambda s).\lambda t \in 2\lambda^2 W$;

hence by subtraction

$(D^2f(a).\lambda t) . \lambda s - (D^2f(a).\lambda s).\lambda t \in 4\lambda^2 W$

for each $|\lambda| \leq \varepsilon_3$. Finally

$(D^2f(a).t).s - (D^2f(a).s).t = f''(a).(t,s) - f''(a).(s,t) \in 4 W$.

This proves the symmetry, since E_2^0 is separated.

9.2. $f^{(p)}$ for $p \geq 1$.

By induction on p we define p-times differentiable mappings as follows : A map $f: E_1 \longrightarrow E_2$ of equable and admissible vector spaces is called p-times differentiable at a, iff $f^{(p-1)}: E_1 \longrightarrow L^*_{p-1}(E_1;E_2)$ exists in a E_1-neighborhood of a and is differentiable at a. Then by definition $D(f^{(p-1)})(a) \in L^*(E_1;L^*_{p-1}(E_1;E_2))$. By (6.4.12) this vector space is linearly homeomorphic to $L^*_p(E_1;E_2)$. We write $f^{(p)}(a)$ for the p-linear mapping thus corresponding to $Df^{(p-1)}(a)$.

(9.2.1) Proposition. If $f : E_1 \longrightarrow E_2$ is p-times differentiable at a point a, then for any fixed $(s_2,s_3,\ldots,s_p)$, the derivative (at a) of the mapping $x \longmapsto f^{(p-1)}(x).(s_2,s_3,\ldots,s_p)$ of E_1 into E_2 is the linear map $s_1 \longrightarrow f^{(p)}(a).(s_1,s_2,\ldots,s_p)$.

Proof. Let $u \in L_n(E_1;E_2)$. Then the mapping $u \longmapsto u(t_1,t_2,\ldots,t_n)$, where $(t_1,\ldots,t_n)$ is fixed, is linear and continuous by (6.2.2). The result follows from (3.3.1), (4.2.1) and the equality $(Df^{(p-1)}(a).s_1) \cdot (s_2,\ldots,s_p) = f^{(p)}(a) \cdot (s_1,s_2,\ldots,s_p)$ (see(6.4.12)).

(9.2.2) Proposition. If $f\colon E_1 \longrightarrow E_2$ is p-times differentiable at $a \in E_1$, then the p-linear mapping $f^{(p)}(a)$ is symmetric.

This is a consequence of the two preceding propositions. The induction proof can be found e.g. in [3] p.177.

(9.2.3) Proposition. If $f\colon E_1 \longrightarrow E_2$ is p-times differentiable and $f^{(p)}\colon E_1 \longrightarrow L_p^{*}(E_1;E_2)$ is q-times differentiable, then $f\colon E_1 \longrightarrow E_2$ is (p+q)-times differentiable.

Proof. $f^{(p)}\colon E_1 \longrightarrow L_p^{*}(E_1;E_2)$ being q-times differentiable, $(f^{(p)})^{(q-1)}\colon E_1 \longrightarrow L_{q-1}^{*}(E_1;L_p^{*}(E_1;E_2))$ is differentiable. Hence by (6.412) also $f^{(p+q-1)}\colon E_1 \longrightarrow L_{p+q-1}^{*}(E_1;E_2)$ is differentiable, which ends the proof.

We call a mapping $f : E_1 \longrightarrow E_2$ infinitely differentiable, iff it is p-times differentiable for all $p \in \mathbf{N}$.

(9.2.4) Example. If $u : E_1 x E_2 \longrightarrow E_3$ is bilinear and continuous, then u is infinitely differentiable.

We already know (see 4.2.3) that Du exists throughout $E_1 x E_2$ and is linear.

To show that $Du : E_1 \times E_2 \longrightarrow L^*(E_1 \times E_2; E_3)$ is continuous, let $\mathfrak{X}_1 \times \mathfrak{X}_2 \downarrow E_1 \times E_2$, $W(\mathcal{A}_1 \times \mathcal{A}_2) \downarrow E_1 \times E_2$. Then

$Du(\mathfrak{X}_1, \mathfrak{X}_2) \cdot (\mathcal{A}_1, \mathcal{A}_2) = u(\mathfrak{X}_1, \mathcal{A}_2) + u(\mathcal{A}_1, \mathfrak{X}_2) \downarrow E_3$

by (9.1.1), (2.9.2), (2.8.8) and (2.3.7).

Hence by (4.2.1) $Du : E_1 \times E_2 \longrightarrow L^*(E_1 \times E_2; E_3)$ is differentiable, and $(Du)'(x_1, x_2) = Du$ for each $(x_1, x_2) \in E_1 \times E_2$. Therefore $u'' : E_1 \times E_2 \longrightarrow L_2^*(E_1 \times E_2; E_3)$ is a constant mapping and thus differentiable by (3.2.3). The same is true for $u^{(k)}$, $k > 2$, which ends the proof.

(9.2.5) Proposition. If $f_1 : E \longrightarrow E_1$ and $f_2 : E \longrightarrow E_2$ are p-times differentiable at a, then the mapping $[f_1, f_2] : E \longrightarrow E_1 \times E_2$ is p-times differentiable at a, and

$$[f_1, f_2]^{(p)}(a) = [f_1^{(p)}(a), f_2^{(p)}(a)].$$

Proof. For $p = 1$ this is (4.4.5). Suppose (9.2.5) for p and let f_1, f_2 be (p+1)-times differentiable at a. We have to prove that $x \longmapsto [f_1, f_2]^{(p)}(x)$ is differentiable at a.

By the induction hypothesis we have $[f_1, f_2]^{(p)}(x) = [f_1^{(p)}(x), f_2^{(p)}(x)]$, which is an element of $L_p(E; E_1 \times E_2)$. This space is linearly homeomorphic to $L_p(E; E_1) \times L_p(E; E_2)$ by (6.4.13), the corresponding element being the pair $(f_1^{(p)}(x), f_2^{(p)}(x))$.

It is therefore sufficient to prove that $x \longmapsto (f_1^{(p)}(x), f_2^{(p)}(x))$ is differentiable, which is now a consequence of (4.4.5), because $(f_1^{(p)}(x), f_2^{(p)}(x)) = [f_1^{(p)}, f_2^{(p)}](x)$ by (1.3.2), and $[f_1^{(p)}, f_2^{(p)}]'(x) = [(f_1^{(p)})'(x), (f_2^{(p)})'(x)]$ corresponds to $[f_1^{(p+1)}(x), f_2^{(p+1)}(x)]$.

(9.2.6) **Theorem.** If $f : E_1 \longrightarrow E_2$ is p-times differentiable at a, and $g : E_2 \longrightarrow E_3$ is p-times differentiable at $b = f(a)$, then $g \circ f : E_1 \longrightarrow E_3$ is p-times differentiable at a.

Proof. For $p = 1$ this is the chain rule (3.3.1). Suppose (9.2.6) proved for p, and let f and g be (p+1)-times differentiable. By (9.2.3), $g \circ f$ is (p+1)-times differentiable, if we show that $(g \circ f)'$ is p-times differentiable. By (1.3.2) and (3.3.1) we can write $(g \circ f)' = c \circ [f', g' \circ f]$, where the bilinear map $c : L^{\#}(E_1;E_2) \times L^{\#}(E_2;E_3) \longrightarrow L^{\#}(E_1;E_3)$ is infinitely differentiable by (6.3.3), (9.1.1) and (9.2.4). The mapping $g' \circ f$ is p-times differentiable by the induction hypothesis. Further by (9.2.5), the same is true for $[f', g' \circ f]$. Finally, applying the induction hypothesis to the maps $[f', g' \circ f]$ and c, one completes the proof.

§ 10. C_k-MAPPINGS.

In § 2 the vector space $C_o(E_1;E_2)$ was introduced, and in § 6 endowed with a pseudo-topology. We recall the definition:

$f \in C_o(E_1;E_2)$ iff :

$\forall \mathcal{A}, \mathfrak{X} \downarrow E_1 \Longrightarrow \forall f(\mathcal{A}), \Delta f(\mathcal{A}, \mathfrak{X}) \downarrow E_2;$

$\mathfrak{F} \downarrow C_o(E_1;E_2)$ iff :

$\forall \mathcal{A} \downarrow E_1 \longrightarrow \mathfrak{F}(\mathcal{A}) \downarrow E_2.$

We now introduce, always assuming that E_1, E_2 are equable and admissible (cf. (9.1.1)), a class of mappings, called C_k-mappings; in the case of finite dimensional spaces, these are exactly the k-times continuously differentiable mappings.

10.1. The vector space $C_k(E_1;E_2)$.

(10.1.1) Definition: For any $k \in \mathbb{N}^o$ (*) we call $f: E_1 \longrightarrow E_2$ a C_k-map and write $f \in C_k(E_1;E_2)$, iff f is k-times differentiable in E_1 and $f^{(k)} \in C_o(E_1;L_k^{\#}(E_1;E_2))$.

(10.1.2) Proposition. If E_1 is finite dimensional and E_2 normable, then $f \in C_k(E_1;E_2)$ if and only if f is k-times continuously differentiable.

This follows by (2.8.6), (2.6.1), (2.6.2), (2.5.2) and (2.8.3).

(*) $\mathbb{N}^o = \{0,1,2,\ldots\}$; for k = 0, we define $f^{(o)} = f$ and $L_o^{\#}(E_1;E_2) = E_2$; for k = 1, $f^{(1)} = f'$ and $L_1^{\#}(E_1;E_2) = L^{\#}(E_1;E_2)$.

(10.1.3) **Proposition.** If $f \in C_k(E_1;E_2)$, then $f^{(n)} \in C_o(E_1;L_n^*(E_1;E_2))$ for $n = 0,\dots,k$.

Proof. The case $k = 0$ is trivial. We prove in detail the case $k = 1$. We then know that $f' \in C_o(E_1;L^*(E_1;E_2))$ and we have to show that $f \in C_o(E_1;E_2)$. So let $\mathbb{V}\mathcal{A} \downarrow E_1$. Let us put $f_o(x) = f(x) - f(o)$ and $\mathcal{B} = [0,1] . \mathcal{A}$. Since $f_o' = f'$ is quasi-bounded and $\mathbb{V}.\mathcal{B} = \mathbb{V}.\mathcal{A} \downarrow E_1$, $\mathbb{V}.f_o'(\mathcal{B}) \downarrow L^*(E_1;E_2)$. Hence, using (2.6.3)

$$(\mathbb{V}.f_o'(\mathcal{B}))(\mathcal{B}) = (\mathbb{V}.f_o)'(\mathcal{B}).\mathcal{B} \downarrow E_2.$$

Since by (1.5.2) and (5.3.6) we have,

$$\mathbb{V}.f(\mathcal{A}) \leq \mathbb{V}.f_o(\mathcal{A}) + \mathbb{V}.f(o) \leq (\mathbb{V}.f_o)(\mathcal{B}) + \mathbb{V}.f(o)$$
$$\leq ((\mathbb{V}f_o)'(\mathcal{B}).\mathcal{B})^{o-} + \mathbb{V}.f(o)$$

we conclude, E_2 being admissible, that $\mathbb{V}.f(\mathcal{A}) \downarrow E_2$. This shows that f is quasi-bounded; it remains to show that it is equably continuous. So let $\mathbb{V}\mathcal{A}, \mathcal{X} \downarrow E_1$.

Since $\Delta f(a,x) = f'(a).x + Rf(a).x$, (1.5.2) yields

$$\Delta f(\mathcal{A},\mathcal{X}) \leq f'(\mathcal{A}).\mathcal{X} + Rf(\mathcal{A}).\mathcal{X} \ .$$

Here, $f'(\mathcal{A}).\mathcal{X} \downarrow E_2$ by (6.2.2), (9.1.1), (2.9.2) and (2.8.8), since $\mathbb{V}.f'(\mathcal{A}) \downarrow L(E_1;E_2)$. The second term is estimated by means of (5.3.6), which can be applied to the filter $Rf(\mathcal{A}) = \mathcal{F}$ since $(Rf(a))(o) = 0$ for all $a \in E_1$. We thus obtain, using that $(Rf(a))'(x) = \Delta f'(a,x)$ and that $\mathcal{X} \leq \mathcal{Y} = \mathbb{V}\mathcal{Y} = [0,1] . \mathcal{Y} \downarrow E_1$

by the equability of E_1 :

$$(Rf(\mathcal{A}))(\mathfrak{X}) \leq (Rf(\mathcal{A})).\mathcal{Y} \leq (\Delta f'(\mathcal{A},\mathcal{Y})(\mathcal{Y}))^{\circ -}.$$

The equable continuity of f' and the admissibility of E_2 implie that the right hand side converges to zero on E_2. This completes the proof for k = 1.

If k > 1, we deduce from $f^{(k)} \in C_o(E_1;L_k^{\#}(E_1;E_2))$ by means of the linear homeomorphisms (6.4.12) and (6.4.17) that $(f^{(k-1)})' \in C_o\big(E_1;L^{\#}(E_1;L_{k-1}^{\#}(E_1;E_2))\big)$, and the same arguments used before yield $f^{(k-1)} \in C_o(E_1;L_{k-1}^{\#}(E_1;E_2))$. Repeating this k-times one completes the proof (see also (7.4.6) and (7.2.3)).

(10.1.4) <u>Corollary</u>. $\underline{C}_{k+n}(E_1;E_2) \subset \underline{C}_k(E_1;E_2)$.

For k > 1, the proof is the same in view of the linear homeomorphism $C_o(E_1;L_k^{\#}(E_1;E_2)) \approx C_o\big(E_1;L^{\#}(E_1;L_{k-1}^{\#}(E_1;E_2))\big)$ where $(f^{(k-1)})'$ corresponds to $f^{(k)}$ and $L_{k-1}^{\#}(E_1;E_2)$ is an admissible vector space. For details see (6.4.12), (6.4.17), (7.4.6) and (7.2.3).

10.2. The structure of $C_k(E_1;E_2)$.

(10.2.1) <u>Definition</u>. $\mathfrak{F} \downarrow C_k(E_1;E_2)$ iff :

$$\mathfrak{F}^{(n)} \downarrow C_o(E_1;L_n^{\#}(E_1;E_2)) \text{ for } n = 0,1,2,\ldots,k.$$

We remark that by (10.1.3) $\mathfrak{F}^{(n)}$ is in fact a filter on the indicated space.

(10.2.2) Lemma. The inclusion (cf. (10.1.4))

$C_{k+n}(E_1;E_2) \subset C_k(E_1;E_2)$ is continuous.

This follows at once by (10.2.1).

(10.2.3) Lemma. The mapping $f \longmapsto f^{(n)}$ of $C_{k+n}(E_1;E_2)$ into $C_k(E_1;L_n^*(E_1;E_2))$ is linear and continuous for each $n \in \mathbb{N}^0$.

Proof. By (6.4.12) and (6.4.17), the corresponding element of $f^{(n+k)} \in C_o(E_1;L_{n+k}^*(E_1;E_2))$ is $(f^{(n)})^{(k)} \in C_o\big(E_1;L_k^*(E_1;L_n^*(E_1;E_2))\big)$, hence $f^{(n)} \in C_k(E_1;L_n^*(E_1;E_2))$. Let now $\mathfrak{F} \downarrow C_{k+n}(E_1;E_2)$. Then by definition $\mathfrak{F}^{(p)} \downarrow C_o(E_1;L_p^*(E_1;E_2))$ for $p = 0,1,\ldots,k+n$.

By (6.4.17) and (6.4.12) we have

$$C_o(E_1;L_p^*(E_1;E_2)) \approx C_o\big(E_1;L_{p-n}^*(E_1;L_n^*(E_1;E_2))\big), \text{ for}$$

$p = n,n+1,\ldots,n+k$. Hence $(\mathfrak{F}^{(n)})^{(p-n)} \downarrow C_o\big(E_1;L_{p-n}^*(E_1;L_n^*(E_1;E_2))\big)$ for $p = n,n+1,\ldots,n+k$, which shows that $\mathfrak{F}^{(n)} \downarrow C_k(E_1;L_n^*(E_1;E_2))$. This proves the continuity.

(10.2.4) Proposition. $C_k(E_1;E_2)$ is an admissible vector space.

Proof. We have to verify that $C_k(E_1;E_2)$ satisfies the compatibility conditions (2.4.2) and the admissibility conditions (cf. (7.1.1)). By (9.1.1), (7.4.6) and (7.2.3), $C_o(E_1;L_n^*(E_1;E_2))$ is admissible for $n = 0,\ldots,k$.

Observing that the mapping $f \longmapsto f^{(n)}$ is linear, one easily verifies the compatibility conditions for $C_k(E_1;E_2)$. We next verify the three admissibility conditions.

a) The inclusion $i : C_k^o(E_1;E_2) \longrightarrow C_o^o(E_1;E_2)$ is continuous by (10.2.4) and (2.9.1). Let $f \in C_k(E_1;E_2)$, $f \neq 0$. Then $f \in C_o(E_1;E_2)$, and by (7.4.6) there is a convex neighborhood V of $o \in C_o^o(E_1;E_2)$ with $f \notin V$. Hence by continuity $i^{-1}(V)$ is a convex neighborhood of $o \in C_k^o(E_1;E_2)$, and $f \notin i^{-1}(V)$. Hence $C_k^o(E_1;E_2)$ is separated.

b) Let $\mathfrak{F} \downarrow C_k(E_1;E_2)$. By (10.2.1) and (7.4.6) we have $\overline{\mathfrak{F}^{(n)}} \downarrow C_o(E_1;L_n^{\#}(E_1;E_2))$, $n = 0,\ldots,k$. But continuity of the map $f \longmapsto f^{(n)}$ ((10.2.3)) implies that $(\overline{\mathfrak{F}})^{(n)} \leq \overline{\mathfrak{F}^{(n)}}$, because $(\overline{F})^{(n)} \subset \overline{F^{(n)}}$ for each $F \in \mathfrak{F}$ (see also (2.9.1)). Hence $\overline{\mathfrak{F}} \downarrow C_k(E_1;E_2)$.

c) Let $\mathfrak{F} \downarrow C_k(E_1;E_2)$. By the linearity of the mapping $f \longmapsto f^{(n)}$, it follows that $(\hat{\mathfrak{F}})^{(n)} = \widehat{\mathfrak{F}^{(n)}}$. Hence $\hat{\mathfrak{F}} \downarrow C_k(E_1;E_2)$.

(10.2.5) Remark. It seems that in general $C_k(E_1;E_2)$ is not equable, not even if E_1,E_2 are topological. For this reason we later on consider $C_k^{\#}(E_1;E_2)$, if the equability condition is required.

Combining (6.1.8) and (10.1.2) we obtain the following generalization of (6.1.9)

(10.2.6) Proposition. If E_1 is a finite dimensional vector space with its natural topology and E_2 a normed vector space, then $C_k(E_1;E_2)$ is the space of k-times continuously differentiable maps from E_1 into E_2 with the topology of uniform convergence on bounded sets of the derivatives of order $0,1,\ldots,k$.

10.3. $C_\infty(E_1;E_2)$.

The admissible vector spaces $C_k(E_1;E_2)$ furnish a projective system of pseudo-topological vector spaces (cf. (2.3.5)), where $\{I, \geqslant\} = \{\mathbb{N}^0, \geqslant\}$ and the inclusion maps are continuous by (10.1.4). We define $C_\infty(E_1;E_2)$ to be the projective limit of this system. Hence by (2.3.5) and (7.3.2)

(10.3.1) $$C_\infty(E_1;E_2) = \bigcap_{k=0}^{\infty} C_k(E_1;E_2)$$

(10.3.2) $\mathfrak{F} \downarrow C_\infty(E_1;E_2)$ iff

$\mathfrak{F} \downarrow C_k(E_1;E_2)$ for each $k \in \mathbb{N}^0$.

(10.3.3) Remark. It is readily verified that the propositions (10.1.2), (10.1.3), (10.2.2) and (10.2.3) are also true for $k = \infty$.

(10.3.4) Proposition. If E_1 is a finite dimensional vector space with its natural topology and E_2 a normed vector space, then $C_\infty(E_1;E_2)$ is the topological projective limit of the topological vector spaces $C_k(E_1;E_2)$.

Proof. In view of (10.2.6), it is sufficient to show that the pseudo-topological projective limit of any projective system of topological vector spaces E_i, $i \in I$, is the same as the topological projective limit, which means that the coarsest pseudo-topology on $E = \bigcap_{i \in I} E_i$ for which the inclusions $f_i : E \longrightarrow E_i$ are continuous, is a topology (*). By (2.3.3) and since $\mathfrak{X} \downarrow E \Longrightarrow f_i(\mathfrak{X}) \downarrow E_i$, we have :

$$f_i(\sup_{\mathfrak{X}\downarrow E} \mathfrak{X}) = \sup_{\mathfrak{X}\downarrow E} (f_i(\mathfrak{X})) \leq \sup_{\mathfrak{Y}\downarrow E_i} \mathfrak{Y} \downarrow E_i .$$

Hence $\sup_{\mathfrak{X}\downarrow E} \mathfrak{X} \downarrow E$, which by (2.4.4) proves the assertion.

10.4. Higher order chain rule.

In order to prove the chain rule for C_k-mappings ((10.4.7)), we need some auxiliary results.

(10.4.1) Proposition. Let $f: E_1 \longrightarrow E_2$ be C_k. Then :

f is $C_{k+p} \Longleftrightarrow f^{(k)}$ is C_p.

(*) For inductive limits, the situation is different: for topological vector spaces, one has to distinguish between the topological inductive limit and the pseudo-topological inductive limit, the latter not necessarily being topological.

Proof. Let $f \in C_{k+p}(E_1;E_2)$. Then f is (k+p)-times differentiable and from $f^{(k+p)} \in C_o(E_1;L^{*}_{k+p}(E_1;E_2))$ we deduce by the linear homeomorphisms (6.4.12) and (6.4.17), that

$$(f^{(k)})^{(p)} \in C_o\Big(E_1;L_p^{*}(E_1;L_k^{*}(E_1;E_2))\Big)$$

and that $f^{(k)}: E_1 \longrightarrow L_k^{*}(E_1;E_2)$ is p-times differentiable. Conversely, if $f^{(k)}$ is C_p, then f is C_{k+p} for the same reasons and by the assumption.

(10.4.2) Proposition. Let $\mathfrak{F}$ be a filter on $C_{k+p+1}(E_1;E_2)$. Then:

$$\mathfrak{F} \downarrow C_{k+p+1}(E_1;E_2) \Longleftrightarrow \Big\{\mathfrak{F}\downarrow C_p(E_1;E_2)\,(*) \text{ and } \mathfrak{F}^{(p+1)} \downarrow C_k(E_1;L^{*}_{p+1}(E_1;E_2))\Big\}.$$

Proof. Let $\mathfrak{F} \downarrow C_{k+p+1}(E_1;E_2)$. Then $\mathfrak{F} \downarrow C_p(E_1;E_2)$ by (10.2.2) and $\mathfrak{F}^{(p+1)} \downarrow C_k(E_1;L_{p+1}^{*}(E_1;E_2))$ by (10.2.3). Conversely, $\mathfrak{F} \downarrow C_p(E_1;E_2)$ means that $\mathfrak{F}^{(m)} \downarrow C_o(E_1;L_m^{*}(E_1;E_2))$ for m = 0,1,...,p and $\mathfrak{F}^{(p+1)} \downarrow C_k(E_1;L_{p+1}^{*}(E_1;E_2))$ implies $\mathfrak{F}^{(p+1+n)} \downarrow C_o(E_1;L^{*}_{p+1+n}(E_1;E_2))$ for n = 0,1,...,k by (6.4.12) and (6.4.17), which ends the proof.

(10.4.3) Proposition. If $\ell \in L(E_1;E_2)$, then $\ell_* : C_k(E;E_1) \longrightarrow C_k(E;E_2)$ defined by $\ell_*(f) = \ell \circ f$ is linear and continuous.

Proof. For k = 0 this is (6.4.16). Suppose the assertion for k and let $f \in C_{k+1}(E;E_1)$. First we show that $\ell \circ f \in C_{k+1}(E;E_2)$. By (3.3.1) we have $(\ell \circ f)'(x) = \ell \circ f'(x)$. The linear mapping $\ell_* : L^{*}(E;E_1) \longrightarrow L^{*}(E;E_2)$ is continuous by (6.3.3).

(*) To be exact, one should write: $i(\mathfrak{F}) \downarrow C_p(E_1;E_2)$, where $i : C_{k+p+1}(E_1;E_2) \longrightarrow C_p(E_1;E_2)$ is the inclusion.

Further $f' \in C_k(E;L^*(E;E_1))$ by (10.2.3). Applying the induction hypothesis to ℓ_* and f', we get

$$(\ell \circ f)' = \ell_* \circ f' \in C_k(E;L^*(E;E_2)),$$

hence $((\ell \circ f)')^{(k)} \in C_o\left(E;L_k^*(E;L^*(E;E_2))\right)$ and $(\ell \circ f)^{(k+1)} \in C_o(E;L_{k+1}^*(E;E_2))$ by the linear homeomorphisms (6.4.12) and (6.4.17).

Let $\mathcal{F} \downarrow C_{k+1}(E;E_1)$. Then $\ell \circ \mathcal{F}$ is a filter on $C_{k+1}(E;E_2)$ by the preceding result $\ell \circ \mathcal{F} \downarrow C_o(E;E_2)$ by (6.4.16) and (10.2.2); $(\ell \circ \mathcal{F})' = \ell_* \circ \mathcal{F}' \downarrow C_k(E;L^*(E;E_2))$, because $\mathcal{F}' \downarrow C_k(E;L^*(E;E_2))$ by (10.2.3) and $g \longmapsto \ell_* \circ g = (\ell_*)_*(g)$ of $C_k(E;L^*(E;E_1))$ into $C_k(E;L^*(E;E_2))$ is continuous by the induction hypothesis. The induction proof is completed by (10.4.2).

(10.4.4) <u>Corollary</u>. If $\ell : E_1 \longrightarrow E_2$ is a linear homeomorphism, then so is $\ell_* : C_k(E;E_1) \longrightarrow C_k(E;E_2)$.

(10.4.5) <u>Proposition</u>. The mapping

$\alpha : C_k(E;E_1) \times C_k(E;E_2) \longrightarrow C_k(E;E_1 \times E_2)$ defined by

$\alpha(f_1,f_2) = [f_1,f_2]$ (see (1.3.1)) is a linear homeomorphism.

<u>Proof</u>. $[f_1,f_2]$ is k-times differentiable by (9.2.5). By definition (10.1.1): $(f_1^{(k)}, f_2^{(k)}) \in C_o(E;L_k^*(E;E_1)) \times C_o(E;L_k^*(E;E_2))$. By (6.4.13) this space is linearly homeomorphic to $C_o(E;L_k^*(E;E_1) \times L_k^*(E;E_2))$, hence also to $C_o(E;L_k^*(E;E_1 \times E_2))$ by (10.4.3), the corresponding element being $[f_1,f_2]^{(k)}$.

In order to prove the continuity of α, one uses the same homeomorphisms and (6.4.14). The continuity of α^{-1} follows from (10.4.3), since $\alpha^{-1}(f) = (\pi_1 \circ f, \pi_2 \circ f)$.

(10.4.6) <u>Proposition</u>. If $u : E_1 \times E_2 \longrightarrow E_3$ is bilinear and continuous, then u is C_∞.

<u>Proof</u>. By (2.8.10) and (9.1.1), u is C_0. By (9.2.4), u is infinitely differentiable. u' is linear and continuous, hence C_0 by (2.8.7). $u^{(k)}$ is a constant map for $k \geq 2$, hence obviously C_0.

(10.4.7) <u>Theorem</u>. If $f \in C_k(E_1;E_2)$ and $g \in C_k(E_2;E_3)$, then $g \circ f \in C_k(E_1;E_3)$.

<u>Proof</u>. For k = 0 this is (2.8.5). Suppose the theorem for k and let f and g be C_{k+1}. By (9.2.6), $g \circ f$ is (k+1)-times differentiable. We assert that

(10.4.8) $$(g \circ f)' \in C_k(E_1;L^*(E_1;E_3)).$$

By (3.3.1) we get $(g \circ f)' = c \circ [f', g' \circ f]$, where $c : L^*(E_1;E_2) \times L^*(E_2;E_3) \longrightarrow L^*(E_1;E_3)$ is C_∞ by (6.3.3) and (10.4.6). $g' \in C_k(E_2;L^*(E_2;E_3))$ and $f' \in C_k(E_1;L^*(E_1;E_2))$ by (10.2.3). $f \in C_k(E_1;E_2)$ by (10.1.4). Applying the induction hypothesis to f and g', we get $g' \circ f \in C_k(E_1;L^*(E_2;E_3))$. Hence $[f', g' \circ f] \in C_k(E_1;L^*(E_1;E_2) \times L^*(E_2;E_3))$ by (10.4.5).

Applying once again the induction hypothesis to c and $[f',g'\circ f]$, we get (10.4.8). From this we conclude $((g\circ f)')^{(k)} \in C_o\big(E_1;L_k{}^{*}(E_1;L^{*}(E_1;E_2))\big)$, hence $(g\circ f)^{(k+1)} \in C_o(E_1;L_{k+1}{}^{*}(E_1;E_2))$ by the linear homeomorphisms (6.4.12) and (6.4.17). One could establish a formula expressing $(g\circ f)^{(k)}$ by means of the derivatives of f and g; the formula is the same as in the classical theory.

§ 11. THE COMPOSITION OF C_k-MAPPINGS.

It will be shown that the composition map $c : C_\infty^{\#}(E_1;E_2) \times C_\infty^{\#}(E_2;E_3) \longrightarrow C_\infty^{\#}(E_1;E_3)$ is not only continuous (*), but even C_∞. We thus get a non trivial example of a C_∞-mapping between spaces which in general are infinite dimensional and not topological. Since the notion of a C_k-mapping was only defined for maps between equable spaces, we have to consider the spaces $C_\infty^{\#}(E_i,E_j)$ (cf.(10.2.5) and (2.6.4)). They coincide with the spaces $C_\infty(E_i;E_j)$ if e.g. the spaces E_1,E_2,E_3 are finite dimensional (cf.(10.3.4)).

11.1. The continuity of the composition map.

(11.1.1) Proposition. Let $u \in L(E_1,E_2;E_3)$. Then the bilinear mapping $\tilde{u} : C_k(E;E_1) \times C_k(E;E_2) \longrightarrow C_k(E;E_3)$ defined by $\tilde{u}(f,g) = u \circ [f,g]$ is equably continuous.

Proof. By (10.4.5), (10.4.6) and (10.4.7), $\tilde{u}(f,g) \in C_k(E;E_3)$. To prove the continuity of $\tilde{u}$, we show

(1) $\mathfrak{F} \downarrow C_k(E;E_1)$, $\mathbb{V}\mathfrak{g} \downarrow C_k(E;E_2) \Longrightarrow \tilde{u}(\mathfrak{F},\mathfrak{g}) \downarrow C_k(E;E_3)$.

(2) $\mathbb{V}\mathfrak{F} \downarrow C_k(E;E_1)$, $\mathfrak{g} \downarrow C_k(E;E_2) \Longrightarrow \tilde{u}(\mathfrak{F},\mathfrak{g}) \downarrow C_k(E;E_3)$.

(*) This continuity statement is not easily comparable with the continuity result of (11.1.4); however the latter is used in order to obtain the differentiability of c.

which will end the proof by (2.8.8). Let $k = 0$, $\mathbb{V}\mathcal{A}\downarrow E$. Then by (1.5.2) and (2.8.8) we have in both cases

$(\tilde{u}(\mathcal{F},\mathcal{G}))(\mathcal{A}) \leq u\,(\mathcal{F}(\mathcal{A}),\mathcal{G}\,(\mathcal{A}))\downarrow E_3$ because either

$\mathcal{F}(\mathcal{A})\downarrow E_1$, $\mathbb{V}\mathcal{G}(\mathcal{A})\downarrow E_2$ or $\mathbb{V}\mathcal{F}(\mathcal{A})\downarrow E_1$, $\mathcal{G}(\mathcal{A})\downarrow E_2$.

Suppose (11.1.1) for k and let us consider the first case:

$$\mathcal{F}\downarrow C_{k+1}(E;E_1),\ \mathbb{V}\mathcal{G}\downarrow C_{k+1}(E;E_2).$$

Then by (10.2.2) and the proof for $k = 0$ it follows that

$\tilde{u}(\mathcal{F},\mathcal{G})\downarrow C_o(E;E_3)$. We assert that

(11.1.2) $$(\tilde{u}(\mathcal{F},\mathcal{G}))'\downarrow C_k(E;L^{\#}(E;E_3))$$

By (6.4.11) and (9.1.1) we have the linear homeomorphisms

(11.1.3) $$L^{\#}(E_1,E_2;E_3) \approx L^{\#}(E_1;L^{\#}(E_2;E_3)).$$

$$L^{\#}(E_1,E_2;E_3) \approx L^{\#}(E_2;L^{\#}(E_1;E_3)).$$

We denote by u_1 resp. u_2 the linear mappings thus corresponding to u. This means $u(x_1,x_2) = u_1(x_1).x_2 = u_2(x_2).x_1$.

It allows us to write $(\tilde{u}(f,g))' = b\circ[g',u_1\circ f] + b\circ[f',u_2\circ g]$ where b is the composition map of linear and continuous mappings, i.e. $b(u,v) = v\circ u$, discussed in (6.3.3), which is bilinear and equably continuous.

By (1.5.2) the above equality yields

$(\tilde{u}(\mathcal{F},\mathcal{G}))' \leq \tilde{b}(\mathcal{G}',u_1\circ\mathcal{F}) + \tilde{b}(\mathcal{F}',u_2\circ\mathcal{G})$

where $\mathbb{V}\mathcal{G}' = (\mathbb{V}\mathcal{G})'\downarrow C_k(E;L^{\#}(E;E_2))$ by (10.2.3) and $u_1 \circ \mathfrak{F} = (u_1)_*(\mathfrak{F}) \downarrow C_k(E;L^{\#}(E_2;E_3))$ by (10.2.2),(11.1.3) and (10.4.3). For the same reasons

$\mathfrak{F}' \downarrow C_k(E;L^{\#}(E;E_1))$ and $\mathbb{V}(u_2 \circ \mathfrak{g}) = u_2 \circ \mathbb{V}\mathfrak{g} \downarrow C_k(E;L^{\#}(E_1;E_3))$.

By (6.3.3),(2.8.8) and the induction hypothesis, applied to the bilinear mapping b, it follows that

$\tilde{b}(\mathfrak{g}', u_1 \circ \mathfrak{F}) \downarrow C_k(E;L^{\#}(E;E_3))$, and $\tilde{b}(\mathfrak{F}', u_2 \circ \mathfrak{g}) \downarrow C_k(E;L^{\#}(E;E_3))$.

Hence the above inequality yields (10.4.9). Applying (10.4.2) for p = 0 this proves that $\tilde{u}(\mathfrak{F}, \mathfrak{g}) \downarrow C_{k+1}(E;E_3)$.

If $\mathbb{V}\mathfrak{F} \downarrow C_{k+1}(E;E_1), \mathfrak{g} \downarrow C_{k+1}(E;E_2)$, then exactly the same proof shows that $\tilde{u}(\mathfrak{F}, \mathfrak{g}) \downarrow C_{k+1}(E;E_3)$.

(11.1.4) Theorem. The mapping

$$c: C_k(E_1;E_2) \times C_k(E_2;E_3) \longrightarrow C_k(E_1;E_3)$$

defined by $c(f,g) = g \circ f$ is continuous.

Proof. We use induction on k. Observe first that

(11.1.5) $$\Delta c((f,g),(\varphi,\psi))(x) = \Delta g(f(x),\varphi(x)) + \psi(f(x)+\varphi(x)).$$

(1) k = 0. Let $\mathfrak{F} \times \mathfrak{g} \downarrow C_o(E_1;E_2) \times C_o(E_2;E_3), \mathbb{V}\mathcal{A} \downarrow E_1$. By (1.5.2), the above equality yields

$$(\Delta c((f,g),(\mathfrak{F},\mathfrak{g})))(\mathcal{A}) \leq \Delta g(f(\mathcal{A}),\mathfrak{F}(\mathcal{A})) + \mathfrak{g}(f(\mathcal{A}) + \mathfrak{F}(\mathcal{A})).$$

By assumption g is C_o, hence $\Delta g(f(\mathcal{A}),\mathfrak{F}(\mathcal{A})) \downarrow E_3$, because $\mathbb{V}f(\mathcal{A}) \downarrow E_2$ and $\mathfrak{F}(\mathcal{A}) \downarrow E_2$. Further $\mathbb{V}(f(\mathcal{A}) + \mathfrak{F}(\mathcal{A})) \leq \mathbb{V}f(\mathcal{A}) + \mathbb{V}\mathfrak{F}(\mathcal{A})$ by (1.5.2), hence also $\mathfrak{g}(f(\mathcal{A}) + \mathfrak{F}(\mathcal{A})) \downarrow E_3$, which ends the proof for k = 0.

(2) Suppose the theorem for k, and let

$$\mathfrak{F} \times \mathfrak{g} \downarrow C_{k+1}(E_1;E_2) \times C_{k+1}(E_2;E_3).$$

We assert that

(11.1.6) $$\Big(\Delta c\big((f,g),(\mathfrak{F},\mathfrak{g})\big)\Big)' \downarrow C_k(E_1;L^{\#}(E_1;E_3)).$$

We again denote by $b : L^{\#}(E_1;E_2) \times L^{\#}(E_2;E_3) \longrightarrow L^{\#}(E_1;E_3)$ the composition map discussed in (6.3.3). Using (3.3.1) and (11.1.1) one gets

(11.1.7) $$\Big(\Delta c\big((f,g),(\varphi,\psi)\big)\Big)' = \tilde{b}\Big(f',\Delta c\big((f,g'),(\varphi,\psi')\big)\Big) + \tilde{b}\Big(\varphi',(g'+\psi')\circ(f+\varphi)\Big)$$

where $\tilde{b}$ is the map associated to b according to (11.1.1) and c in the right hand side expression is the composition map:

$$C_{k+1}(E_1;E_2) \times C_k(E_2;L^{\#}(E_2;E_3)) \longrightarrow C_k(E_1;L^{\#}(E_2;E_3)).$$

Applying the induction hypothesis to this map and using (10.2.2) and (10.2.3), one gets

(11.1.8) $$\Delta c\big((f,g'),(\mathfrak{F},\mathfrak{g}')\big) \downarrow C_k(E_1;L^{\#}(E_2;E_3)).$$

$\mathbb{W}f \downarrow C_{k+1}(E_1;E_2)$ by the compatibility condition (4) of (2.4.2), hence $(\mathbb{W}f)' = \mathbb{W}f' \downarrow C_k(E_1;L^{\#}(E_1;E_2))$ by (10.2.3). $\tilde{b}$ being equably continuous by (11.1.1), this proves that

(11.1.9) $$\tilde{b}\Big(f',\Delta c\big((f,g'),(\mathfrak{F},\mathfrak{g}')\big)\Big) \downarrow C_k(E_1;L^{\#}(E_1;E_3)).$$

Using the equality $(g'+\psi')\circ(f+\varphi) = \Delta c((f,g'),(\varphi,\psi')) + g'\circ f$ and (1.5.2) one gets

$$\mathbb{W}.(g'+\mathfrak{g}')\circ(f+\mathfrak{F}) \leq \mathbb{W}.\Delta c\big((f,g'),(\mathfrak{F},\mathfrak{g}')\big) + \mathbb{W}.g'\circ f \downarrow C_k(E_1;L^{\#}(E_2;E_3))$$

by the compatibility conditions of (2.4.2), by (10.4.7) and (11.1.8).

$$\mathfrak{f}' \downarrow C_k(E_1;L^{\#}(E_1;E_2)) \text{ by (10.2.3), hence also}$$

(11.1.10) $$\tilde{b}\left(\mathfrak{f}',(g'+\mathfrak{g}')\circ(f+\mathfrak{f})\right) \downarrow C_k(E_1;L^{\#}(E_1;E_3)).$$

The equality (11.1.7) yields

$$\left(\Delta c\left((f,g'),(\mathfrak{f},\mathfrak{g}')\right)\right)' \leq \tilde{b}\left(f',\Delta c\left((f,g'),(\mathfrak{f},\mathfrak{g}')\right)\right) + \tilde{b}\left(\mathfrak{f}',(g'+\mathfrak{g}')\circ(f+\mathfrak{f})\right) \downarrow C_k(E_1;L^{\#}(E_1;E_3))$$

by (11.1.9) and (11.1.10), which proves the assertion (11.1.6). The proof for $k = 0$ also shows that

$$\Delta c\left((f,g),(\mathfrak{f},\mathfrak{g})\right) \downarrow C_o(E_1;E_3),$$

hence applying (10.4.2) for $p = 0$, one completes the induction proof.

11.2. The differentiability of the composition map.

In order to prove the general theorem, we establish some auxiliary results.

(11.2.1) Proposition. Let $g \in C_{k+1}(E_2;E_3)$. Then the mapping $g_*: C_k(E_1;E_2) \longrightarrow C_k(E_1;E_3)$ defined by $g_*(f) = g\circ f$ is differentiable throughout $C_k(E_1;E_2)$, and $(g_*)'(f).\varphi = \tilde{e}(g'\circ f,\varphi)$.

<u>Proof.</u> Consider the expression $r_g(\varphi) = g\circ(f+\varphi) - g\circ f - e\circ[g'\circ f, \varphi]$, where $e : L^{\#}(E_2;E_3) \times E_2 \longrightarrow E_3$ is the evaluation map discussed in (6.2.3). According to (11.1.1) we use the notation

$$e\circ [g'\circ f, \varphi] = \tilde{e}(g'\circ f, \varphi).$$

By (11.1.1) the linear mapping $\varphi \longmapsto \tilde{e}(g'\circ f, \varphi)$ is continuous, $g'\circ f$ being an element of $C_k(E_1;L^{\#}(E_2;E_3))$ by (10.4.7) and (10.2.3). The proof is complete if we show that

(11.2.2) $$r_g \in R\big(C_k(E_1;E_2);\ C_k(E_1;E_3)\big).$$

We use induction on k.

(A) Let $k = 0$, $\mathbb{W}\mathcal{A} \downarrow C_o(E_1;E_2)$ and $\mathbb{W}\mathcal{B} \downarrow E_1$. Then by (5.3.5), (9.1.1), (1.5.2) and using the equality

(11.2.3) $$\begin{aligned} \Theta r_g(\lambda, \varphi).x &= \frac{1}{\lambda}\Big(g\big(f(x)+\lambda\varphi(x)\big) - g(f(x)) - \\ &\quad g'(f(x)).\lambda\varphi(x)\Big) \\ &= \Theta Rg\big(f(x)\big) . \big(\lambda, \varphi(x)\big), \end{aligned}$$

we get $\big(\Theta r_g(\mathbb{W}, \mathcal{A})\big)(\mathcal{B}) \leq \Theta Rg\big(f(\mathcal{B})\big) . \big(\mathbb{W}, \mathcal{A}(\mathcal{B})\big)$

$$\leq \Big(\Delta g'\big(f(\mathcal{B}), \mathbb{W}\mathcal{A}(\mathcal{B})\big) . \mathcal{A}(\mathcal{B})\Big)^{o-} \downarrow E_3,$$

because $\big(Rg(f(x))\big)'(h) = \Delta g'(f(x),h)$ and $\big(Rg(f(x))\big)(0) = 0$ for each $x,h \in E_1$.

This proves that $\Theta r_g(\mathbb{W}, \mathcal{A}) \downarrow C_o(E_1;E_3)$.

(B) Suppose (11.2.1) for k and let $g \in C_{k+2}(E_2;E_3)$, $\mathbb{W}\mathcal{A} \downarrow C_{k+1}(E_1;E_2)$.

We assert:

(11.2.4) $$\left(\theta r_g(\mathbb{V},\mathcal{A})\right)' \downarrow C_k(E_1;L^{\#}(E_1;E_3)),$$

which will end the proof by (A) and (10.4.2).

Using (11.2.3) and the chain rule one finds that the derivative of the mapping $x \longmapsto \theta r_g(\lambda,\varphi).\,x$ at a point $x \in E_1$ is the linear map

$$s \longmapsto \frac{1}{\lambda}\Big(\big(g'\circ(f+\lambda\varphi)\big)(x) - \big(g'\circ f\big)(x)\Big).f'(x).s$$
$$+ \big(g'(f+\lambda\varphi)\big)(x).\big(\varphi'(x).s\big) - g'\big(f(x)\big).\big(\varphi'(x).s\big)$$
$$- g''\big(f(x)\big).\big(f'(x).s,\varphi(x)\big).$$

By (9.1.3) one gets:

$$g''\big(f(x)\big).\big(f'(x).s,\varphi(x)\big) = \big(\tilde{e}(D^2g\circ f,\varphi)(x)\circ f'(x)\big).s, \text{ hence}$$

$$\big(\theta r_g(f).(\lambda,\varphi)\big)'(x) = \frac{1}{\lambda}\ \big(g'\circ(f+\lambda\varphi) - g'\circ f - \tilde{e}(D^2g\circ f,\lambda\varphi)\big)(x)\circ f'(x)$$
$$+ \big(g'\circ(f+\lambda\varphi) - g'\circ f\big)(x)\circ\varphi'(x), \text{ which yields}$$

$$\big(\theta r_g(\lambda,\varphi)\big)' = \tilde{b}\big(f',\theta r_{g'}(\lambda,\varphi)\big) + \tilde{b}\big(\varphi'\,\Delta(g')_*(f,\lambda\varphi)\big),$$

where $(g')_*(f) = g'\circ f$ and

(11.2.5) $$r_{g'}(\varphi) = g'\circ(f+\varphi) - g'\circ f - \tilde{e}\big((g')'\circ f,\varphi\big).$$

By (1.5.2) we obtain therefore

(11.2.6) $$\big(\theta r_g(\mathbb{V},\mathcal{A})\big)' \leq \tilde{b}\big(f',\theta r_{g'}(\mathbb{V},\mathcal{A})\big) + \tilde{b}\big(\mathcal{A}',\Delta(g')_*(f,\mathbb{V}\mathcal{A})\big).$$

Since $g' \in C_{k+1}(E_2;L^{\#}(E_2;E_3))$, the induction hypothesis applied to $(g')_*$ yields : $\theta r_{g'}(\mathbb{V},\mathcal{A}) \downarrow C_k(E_1;L^{\#}(E_2;E_3))$; further by (3.2.5) and (2.6.3):

$$\Delta(g')_*(f,\mathbb{V}\mathcal{A}) \downarrow C_k(E_1;L^{\#}(E_2;E_3)).$$

Applying (11.1.1) to the bilinear map b and using (2.8.8) one proves (11.2.4) by the inequality (11.2.6).

(11.2.7) Proposition. The mapping

$c: C_k(E_1;E_2) \times C_{k+1}(E_2;E_3) \longrightarrow C_k(E_1;E_3)$

is differentiable throughout $C_k(E_1;E_2)$ and

$c'(f,g).(\varphi,\psi) = \tilde{e}(g' \circ f, \varphi) + \psi \circ f$

Proof. We verify the assumptions of (8.2.1). $D_1c(f,g).\varphi = \tilde{e}(g' \circ f, \varphi)$ by (11.2.1). $D_2c(f,g).\psi = \psi \circ f$, because c is linear in the second variable and continuous by (11.1.4).

$C_k(E_1;E_3)$ is an admissible vector space by (10.2.4). The mapping $D_1c : C_k(E_1;E_2) \times C_{k+1}(E_2;E_3) \longrightarrow L(C_k(E_1;E_2);C_k(E_1;E_3))$, is continuous, because $W\mathcal{A} \downarrow C_k(E_1;E_2)$ and $\mathcal{F} \times \mathcal{G} \downarrow C_k(E_1;E_2) \times C_{k+1}(E_2;E_3)$ imply $\left(D_1c(f+\mathcal{F},g+\mathcal{G}) - D_1c(f,g)\right)(\mathcal{A}) = \tilde{e}\left(\Delta c((f,g'),\mathcal{F} \times \mathcal{G}'),\mathcal{A}\right)$ $\downarrow C_k(E_1;E_3)$ by (11.1.1) and (2.8.8).

(11.2.8) Lemma. If E_2 is an admissible but not necessarily equable vector space, $f : E_1 \longrightarrow E_2$ is differentiable and $f' : E_1^{\#} \longrightarrow L(E_1^{\#};E_2^{\#})$ is continuous, then $f: E_1^{\#} \longrightarrow E_2^{\#}$ is differentiable.

Proof. Let $a \in E_1$. By assumption $f(a+x) = f(a) + \ell(x) + r(x)$ where $\ell \in L(E_1;E_2)$ and $r \in R(E_1;E_2)$. Since by (2.9.1) we have $L(E_1;E_2) \subset L(E_1^{\#};E_2^{\#})$, it only remains to show that $r \in R(E_1^{\#};E_2^{\#})$.

So let $\mathbb{V}\mathcal{B} \downarrow E_1^{*}$. Since $r(0) = 0$ and $r'(x) = \Delta f'(a,x)$, we deduce from (5.3.5) : $\Theta r(\mathbb{V},\mathcal{B}) \leq \left(\Delta f'(a, \mathbb{V}\mathcal{B}).\mathcal{B}\right)^{0-}$ which converges to zero on E_2^{*}, because by the continuity of f' we have $\Delta f'(a, \mathbb{V}\mathcal{B}) \downarrow L(E_1^{*};E_2^{*})$ and because E_2^{*} is admissible (cf.(7.2.3)).

(11.2.9) <u>Lemma</u>. If $\mathbb{V}\mathcal{A} \downarrow C_k(E_1;E_2)$ and $\mathcal{g} \downarrow C_k(E_2;E_3)$, then $\mathcal{g}\circ\mathcal{A} \downarrow C_k(E_1;E_3)$.

<u>Proof</u>. Let $k = 0$, $\mathbb{V}\mathcal{B} \downarrow E_1$. Then $(\mathcal{g}\circ\mathcal{A})(\mathcal{B}) = \mathcal{g}(\mathcal{A}(\mathcal{B})) \downarrow E_3$, because $\mathbb{V}(\mathcal{A}(\mathcal{B})) = (\mathbb{V}\mathcal{A})(\mathcal{B}) \downarrow E_2$. Suppose the lemma for k and let $\mathbb{V}\mathcal{A} \downarrow C_{k+1}(E_1;E_2)$, $\mathcal{g} \downarrow C_{k+1}(E_2;E_3)$. We have $(g\circ f)' = \tilde{b}(f', g'\circ f)$, hence by (1.5.2) $(\mathcal{g}\circ\mathcal{A})' \leq \tilde{b}\,(\mathcal{A}', \mathcal{g}'\circ\mathcal{A})$ which converges to zero on $C_k(E_1;L^{*}(E_1;E_3))$ by (10.2.3), (11.1.1), (2.8.8) and the induction hypothesis. The proof for $k = 0$ also shows that $\mathcal{g}\circ\mathcal{A} \downarrow C_0(E_1;E_3)$, hence the result by (10.4.2). Since $\mathcal{g}$ equable implies $\mathcal{g}\circ\mathcal{A}$ equable, we have :

(11.2.10) <u>Corollary</u>. If $\mathbb{V}\mathcal{A} \downarrow C_k^{*}(E_1;E_2)$ and $\mathcal{g} \downarrow C_k^{*}(E_2;E_3)$, then $\mathcal{g}\circ\mathcal{A} \downarrow C_k^{*}(E_1;E_3)$.

(11.2.11) <u>Lemma</u>. If $\mathcal{g} \downarrow C_{k+1}(E_2;E_3)$ and $\mathbb{V}\mathcal{A}_1, \mathbb{V}\mathcal{A}_2 \downarrow C_k(E_1;E_2)$, then $\Theta R\, \mathcal{g}_*(\mathcal{A}_1).(\mathbb{V},\mathcal{A}_2) \downarrow C_k(E_1;E_3)$.

<u>Proof</u>. The equality $\Delta(g')_*(f,\varphi) = g'\circ(f+\varphi) - g'\circ f$ yields $\Delta(\mathcal{g}')_*(\mathcal{A}_1, \mathbb{V}\mathcal{A}_2) \leq \mathcal{g}'\circ(\mathcal{A}_1 + \mathbb{V}\mathcal{A}_2) - \mathcal{g}'\circ\mathcal{A}_1$ which converges to zero on $C_k(E_1;L^{*}(E_2;E_3))$ by the preceding lemma and (10.2.3).

From the equality (cf.(11.2.1)) $\Delta(g_*)'(f,\varphi_1)\cdot\varphi_2 = \tilde{e}(\Delta(g')_*(f,\varphi_1),\varphi_2)$ one deduces: $\Delta(\mathcal{g}_*)'(\mathcal{A}_1,\mathbb{V}\mathcal{A}_2).\mathcal{A}_2 = \tilde{e}(\Delta(\mathcal{g}')_*(\mathcal{A}_1,\mathbb{V}\mathcal{A}_2),\mathcal{A}_2)$, and this converges to zero on $C_k(E_1;E_3)$ by the preceding statement, (11.1.1) and (2.8.8).

We further have $R\, g_*(f).0 = 0$ and

$$(Rg_*(f))'(\varphi) = \Delta(g_*)'(f,\varphi),$$

hence by (5.3.5) $\theta R\, \mathcal{g}_*(\mathcal{A}_1).(\mathbb{V},\mathcal{A}_2) \leq (\Delta(\mathcal{g}_*)'(\mathcal{A}_1,\mathbb{V}\mathcal{A}_2).\mathcal{A}_2)^{0-}$

which converges to zero on $C_k(E_1;E_3)$ since $C_k(E_1;E_3)$ is admissible by (10.2.4).

(11.2.12) Lemma. If $\mathbb{V}\mathcal{g}\downarrow C_{k+1}(E_2;E_3)$ and $\mathcal{F}$, $\mathbb{V}\mathcal{A}\downarrow C_k^{\#}(E_1;E_2)$, then $R\,\mathcal{g}_*(\mathcal{A}).\mathcal{F}\downarrow C_k^{\#}(E_1;E_3)$.

Proof. One easily verifies the following equalities:

(1) $Rg_*(f).(\lambda\varphi) = \theta R(\lambda g)_*(f).(\lambda,\varphi)$

(2) $\lambda\,\theta Rg_*(f).(\mu,\varphi) = \theta R(\lambda g)_*(f).(\mu,\varphi)$.

If $\mathcal{F}\downarrow C_k^{\#}(E_1;E_2)$, then there is $\mathcal{F}_1$ with $\mathcal{F}\leq\mathbb{V}\mathcal{F}_1 = \mathcal{F}_1$ and $\mathcal{F}_1\downarrow C_k^{\#}(E_1;E_2)$. By (1), (1.5.2) and (11.2.10) we therefore have $R\mathcal{g}_*(\mathcal{A})(\mathcal{F}) \leq R\mathcal{g}_*(\mathcal{A}).(\mathbb{V}\mathcal{F}_1) \leq \theta R(\mathbb{V}\mathcal{g})_*(\mathcal{A}).(\mathbb{V},\mathcal{F}_1)\downarrow C_k(E_1;E_3)$.

By (2) one gets

$\theta R(\mathbb{V}\mathcal{g})_*(\mathcal{A}).(\mathbb{V},\mathcal{F}_1) = \mathbb{V}(\theta R\mathcal{g}_*(\mathcal{A}).(\mathbb{V},\mathcal{F}_1))$ which is an equable filter and thus converges to zero even on $C_k^{\#}(E_1;E_3)$.

(11.2.13) **Lemma.** Let $c_2 : C_k^{\#}(E_1;E_2) \longrightarrow L(C_{k+p+1}^{\#}(E_2;E_3);C_k^{\#}(E_1;E_3))$ be the mapping defined by $c_2(f) = f^*$, where $f^*(\psi) = \psi \circ f$ for any $\psi \in C_{k+p+1}(E_2;E_3)$. Then c_2 is C_0.

Proof. (a) Let $\mathbb{W}\mathcal{A}\downarrow C_k^{\#}(E_1;E_2)$, $\mathbb{W}\mathcal{G}\downarrow C_{k+p+1}^{\#}(E_2;E_3)$. Then $\mathbb{W}c_2(\mathcal{A}).\mathcal{G} = \mathbb{W}\mathcal{G}\circ\mathcal{A}\downarrow C_k^{\#}(E_1;E_3)$ by (11.2.10), which shows that c_2 is quasi-bounded.

(b) Let $\mathbb{W}\mathcal{A},\mathcal{F}\downarrow C_k^{\#}(E_1;E_2)$, $\mathbb{W}\mathcal{G}\downarrow C_{k+p+1}^{\#}(E_1;E_3)$. For any $\psi \in C_{k+p+1}^{\#}(E_2;E_3)$ we have by (11.2.1), (10.1.4) and (2.6.2) :

(11.2.14) $\Delta c_2(f,\varphi).\psi = \psi\circ(f+\varphi) + \psi\circ f = \tilde{a}(\psi'\circ f,\varphi) + R\psi_*(f).\varphi.$

Hence by (1.5.2)

$\Delta c_2(\mathcal{A},\mathcal{F}).\mathcal{G} \leq \tilde{a}(\mathcal{G}'\circ\mathcal{A},\mathcal{F}) + R\mathcal{G}_*(\mathcal{A}).\mathcal{F}$ where $\tilde{a}(\mathcal{G}'\circ\mathcal{A},\mathcal{F})$ $\downarrow C_k^{\#}(E_1;E_3)$ by (6.2.3), (11.1.1), (2.8.8), (10.2.3) and (11.2.10). $R\mathcal{G}_*(\mathcal{A}).\mathcal{F}\downarrow C_k^{\#}(E_1;E_3)$ by the preceding lemma, hence c_2 is equably continuous. (a) and (b) prove the lemma.

(11.2.15) **Lemma.** For any $f,\varphi \in C_k(E_1;E_2)$ $r_f(\varphi) : C_{k+p+2}(E_2;E_3) \longrightarrow C_k(E_1;E_3)$ denotes the linear map defined by

(11.2.16) $r_f(\varphi).\psi = R\psi_*(f).\varphi = \psi\circ(f+\varphi) - \psi\circ f - \tilde{a}(\psi'\circ f,\varphi).$

Then we have: If $\mathbb{W}\mathcal{A}$, $\mathbb{W}\mathcal{F}\downarrow C_k^{\#}(E_1;E_2)$ and $\mathbb{W}\mathcal{G}\downarrow C_{k+p+2}^{\#}(E_2;E_3)$, then $\Theta r_{\mathcal{F}}(\mathbb{V},\mathcal{A}).\mathcal{G}\downarrow C_k^{\#}(E_1;E_3)$.

Proof. One easily verifies the following equalities:

(1) $\Delta(\psi')_*(f,\varphi) = \Delta c_2(f,\varphi).\psi'$;

(2) $\Delta(\psi_*)'(f,\varphi) = \big(R\,\psi_*(f)\big)'(\varphi)$;

(3) $\tilde{e}\big(\Delta(\psi')_*(f,\varphi_1),\varphi_2\big) = \Delta(\psi_*)'(f,\varphi_1).\varphi_2$.

By (10.2.3) we get $(\mathbb{W}g)' = \mathbb{W}g' \downarrow C^{\#}_{k+p+1}\big(E_2;L^{\#}(E_2;E_3)\big)$.

Hence by (11.2.13) and (1) :

$\Delta(g')_*(\mathfrak{F},\mathbb{W}\mathcal{A}) \downarrow C_k^{\#}(E_1;L^{\#}(E_2;E_3))$ and by (3), (6.2.3), (11.1.1), (2.8.8) and (11.2.1):

$\tilde{e}\big(\Delta(g')_*(\mathfrak{F},\mathbb{W}\mathcal{A}),\mathcal{A}\big) = \Delta(g_*)'(\mathfrak{F},\mathbb{W}\mathcal{A}).\mathcal{A} \downarrow C_k^{\#}(E_1.E_3)$.

Using (2), (5.3.5), (10.2.4), (7.2.3), (11.2.15) and the above result we get

$$\Theta r_{\mathfrak{F}}\ (\mathbb{W},\mathcal{A}).g = \Theta R g_*(\mathfrak{F}).(\mathbb{W},\mathcal{A}) \leq \big(\Delta(g_*)'(\mathfrak{F},\mathbb{W}\mathcal{A}).\mathcal{A}\big)^{0-}$$

$\downarrow C_k^{\#}(E_1;E_3)$, hence the assertion of the lemma.

(11.2.17) Lemma. The mapping (cf. (11.2.13))

$$c_2 : C_k^{\#}(E_1;E_2) \longrightarrow L\big(C^{\#}_{k+p+2}(E_2;E_3);C_k^{\#}(E_1;E_3)\big)$$

is differentiable.

Proof. By (11.2.14) we have $\big(c_2(f+\varphi)-c_2(f)\big).\psi = \tilde{e}\big(\psi'\circ f,\varphi\big)+R\psi_*(f).\varphi$.

For any $f \in C_k(E_1;E_2)$

$$\ell_f : C_k^{\#}(E_1;E_2) \longrightarrow L\big(C^{\#}_{k+p+2}(E_2;E_3);C_k^{\#}(E_1;E_3)\big)$$

denotes the mapping caracterized by $\ell_f(\varphi).\psi = \tilde{e}(\psi'\circ f,\varphi)$.

Similarly we had defindd r_f (cf.(11.2.16)) by

$$r_f(\varphi).\psi = R\psi_*(f).\varphi.$$

Obviously ℓ_f is linear; it remains to show that ℓ_f is continuous and that r_f is a remainder.

Let $\mathfrak{F} \downarrow C_k^{\#}(E_1;E_2)$, $\mathbb{V}\mathcal{G} \downarrow C_{k+p+2}^{\#}(E_2;E_3)$. Then

$\ell_f(\mathfrak{F}).\mathcal{G} = \tilde{e}(\mathcal{G}' \circ f, \mathfrak{F}) \downarrow C_k^{\#}(E_1;E_3)$ by (6.2.3), (11.1.1), (2.8.8) and (11.2.10). Obviously $\mathbb{V}f \downarrow C_k^{\#}(E_1;E_2)$, hence by (11.2.15):

$\Theta r_f(\mathbb{V},\mathcal{A}).\mathcal{G} \downarrow C_k^{\#}(E_1;E_3)$, if $\mathbb{V}\mathcal{A} \downarrow C_k^{\#}(E_1;E_2)$, which completes the proof.

(11.2.18) <u>Lemma</u>. The mapping

$$c_2 \colon C_k^{\#}(E_1;E_2) \longrightarrow L^{\#}\left(C_{k+p+2}^{\#}(E_2;E_3);C_k^{\#}(E_1;E_3)\right) \text{ is } \mathcal{C}_o.$$

<u>Proof</u>. (a) Let $\mathbb{V}\mathcal{A} \downarrow C_k^{\#}(E_1;E_2)$. If $\mathbb{V}\mathcal{G} \downarrow C_{k+p+2}^{\#}(E_2;E_3)$, then

$\mathbb{V}.c_2(\mathcal{A}).\mathcal{G} = \mathbb{V}\mathcal{G} \circ \mathcal{A} \downarrow C_k^{\#}(E_1;E_3)$ by (11.2.10). Since furthermore $\mathbb{V}.c_2(\mathcal{A})$ is an equable filter, it follows that c_2 is quasi-bounded.

(b) By the differentiability of c_2 we have

$$\Delta c_2(f,\varphi) = c_2(f+\varphi) - c_2(f) = c_2'(f).\varphi + Rc_2(f).\varphi.$$

Let $\mathbb{V}\mathcal{A},\mathfrak{F} \downarrow C_k^{\#}(E_1;E_2)$. Then by (1.5.2)

(11.2.19) $\Delta c_2(\mathcal{A},\mathfrak{F}) \leq c_2'(\mathcal{A}).\mathfrak{F} + Rc_2(\mathcal{A}).\mathfrak{F}.$

$\mathbb{V}c_2'(\mathcal{A}) \downarrow L\left(C_k^{\#}(E_1;E_2);L\left(C_{k+p+2}^{\#}(E_2;E_3);C_k^{\#}(E_1;E_3)\right)\right)$, because $\mathbb{V}.\mathcal{B}_1 \downarrow C_k^{\#}(E_1;E_2)$, $\mathbb{V}\mathcal{B}_2 \downarrow C_{k+p+2}^{\#}(E_2;E_3)$ implies

$(\mathbb{V}c_2'(\mathcal{A}).\mathcal{B}_1).\mathcal{B}_2 = \tilde{e}\,(\mathcal{B}_2' \circ \mathcal{A}, \mathbb{V}\mathcal{B}_1) \downarrow C_k^{\#}(E_1;E_3)$ by (6.2.3), (11.1.1), (2.8.8) and (11.2.10).

Since we can suppose $\mathbb{W}\mathfrak{F} = \mathfrak{F}$, we get

$$c_2'(\mathcal{A}).\mathfrak{F} = \mathbb{W}.c_2'(\mathcal{A}).\mathfrak{F} \downarrow L^{\#}\left(C^{\#}_{k+p+2}(E_2;E_3);C_k^{\#}(E_1;E_3)\right).$$

The equality $Rc_2(f).(\lambda\varphi) = \lambda.\ \theta Rc_2(f).(\lambda,\varphi)$ yields by (1.5.2)

$$Rc_2(\mathcal{A}).\mathfrak{F} = Rc_2(\mathcal{A}).(\mathbb{W}\mathfrak{F}) \leq \mathbb{W}.\ \theta Rc_2(\mathcal{A}).(\mathbb{W},\mathfrak{F}).$$

This filter being equable, it remains to show that for

$\mathbb{W}\mathcal{G} \downarrow C^{\#}_{k+p+2}(E_2;E_3)$ one has $\left(\theta Rc_2(\mathcal{A}).(\mathbb{W},\mathfrak{F})\right).\mathcal{G} \downarrow C_k^{\#}(E_1;E_3)$.

But this follows at once from (11.2.15), because

$$(Rc_2(f).\varphi).\psi = r_f(\varphi).\psi$$

(cf.(11.2.17)). Now the equable continuity of c_2 follows from the inequality (11.2.19).

(11.2.20) Lemma. The mapping

$$c_2:\ C_k^{\#}(E_1;E_2) \longrightarrow L^{\#}\left(C^{\#}_{k+p+2}(E_2;E_3);\ C_k^{\#}(E_1;E_3)\right) \text{ is } C_p.$$

Proof. For p = 0 this is (11.2.18). Suppose the assertion for p and consider the mapping

$$c_2:\ C_k^{\#}(E_1;E_2) \longrightarrow L^{\#}\left(C^{\#}_{k+p+3}(E_2;E_3);\ C_k^{\#}(E_1;E_3)\right).$$

We make use of the following abbreviations:

$E_4 = C_k^{\#}(E_1;E_2)$;

$E_5 = C^{\#}_{k+p+3}(E_2;E_3)$;

$E_6 = C_k^{\#}(E_1;E_3)$;

$E_7 = C^{\#}_{k+p+2}(E_2;L^{\#}(E_2;E_3))$;

$E_8 = C_k^{\#}(E_1;L^{\#}(E_2;E_3))$.

We assert that the mapping

(A) $c_2' : E_4 \longrightarrow L^{*}(E_4;L^{*}(E_5;E_6))$ is C_p.

Observe that this mapping exists by (11.2.17), (2.9.1) and (6.4.10). The assertion follows from (6.4.11) and (10.4.4), if we show:

(B) The mapping

$$c_{21} : E_4 \longrightarrow L^{*}(E_4, E_5; E_6)$$

defined by $c_{21}(f).(\varphi,\psi) = (c_2'(f).\varphi).\psi = \tilde{e}(\psi'\circ f,\varphi)$ is C_p.

We denote by $\pi_1 : E_4 \times E_5 \longrightarrow E_4$ resp. $\pi_2 : E_4 \times E_5 \longrightarrow E_5$ the projection $(\varphi,\psi) \longmapsto \varphi$ resp. $(\varphi,\psi) \longrightarrow \psi$.

Then we have $c_{21}(f) = \tilde{\tilde{e}} \circ [c_2(f)\circ D\circ \pi_2, \pi_1]$ where

$D : E_5 \longrightarrow E_7$ is linear and continuous by (10.2.3).

$c_2 : E_4 \longrightarrow L^{*}(E_7;E_8)$ is C_p by our induction hypothesis and

$\tilde{e} : E_8 \times E_4 \longrightarrow E_6$ is C_p by (11.1.1) and (10.4.6).

Applying (11.1.1) to $\tilde{e}$ which is an element of $L(E_8, E_4; E_6)$, we conclude that

$\tilde{\tilde{e}} : L^{*}(E_4\times E_5;E_4) \times L^{*}(E_4\times E_5;E_4) \longrightarrow L^{*}(E_4\times E_5;E_6)$

is bilinear and continuous by (2.3.6). One has of course to verify that for any $k \in \mathbb{N}$ $L^{*}(E_1;E_2)$ has the structure induced by its inclusion in $C_k^{*}(E_1;E_2)$.

Let $\tau_1 : E_4 \longrightarrow L^{\#}(E_4 \times E_5; E_8)$ resp. $\tau_2 : E_4 \longrightarrow L^{\#}(E_4 \times E_5; E_4)$ be the mappings defined by $\tau_1(f) = c_2(f) \circ D \circ \pi_2$ resp. $\tau_2(f) = \pi_1 \cdot \tau_2$ is a constant map, hence obviously C_p. $\tau_1 = (D \circ \pi_2)^{*} \circ c_2$ is the composite of C_p mappings hence also C_p by (10.4.7). We have therefore: $c_{21} = \tilde{e} \circ [\tau_1, \tau_2]$, and this is the composite of C_p-mappings by (10.4.5), which proves (B) and hence (A). By (11.2.17) we know that

$c_2 : E_4 \longrightarrow L(E_5; E_6)$ is differentiable.

By (A), (2.6.3) and (2.8.3), $c_2' : E_4 \longrightarrow L(E_4; L^{\#}(E_5; E_6))$ is continuous. $L(E_5; E_6)$ is an admissible vector space by (7.4.6). Hence from (11.2.8) it follows that $c_2 : E_4 \longrightarrow L^{\#}(E_5; E_6)$ is differentiable. One deduces from (A) that c_2 is C_1 and c_2' is C_p and thus by (10.4.1) c_2 is C_{p+1} which ends our induction proof of (11.2.20).

(11.2.21) Theorem. The mapping

$$c: C_k^{\#}(E_1; E_2) \times C_{k+p+1}^{\#}(E_2; E_3) \longrightarrow C_k^{\#}(E_1; E_3) \text{ is } C_p.$$

Proof. (A) $p = 0$. By (11.2.13) we have

$$\Delta c((f,g),(\varphi,\psi)) = \Delta c_2(f,\varphi).g + \psi \circ (f+\varphi).$$

Let $\forall \mathcal{A}, \mathcal{F} \downarrow C_k^{\#}(E_1; E_2)$ and $\forall \mathcal{B}, \mathcal{G} \downarrow C_{k+1}^{\#}(E_2; E_3)$. Then by (1.5.2)
$\Delta c(\mathcal{A} \times \mathcal{B}, \mathcal{F} \times \mathcal{G}) \leq \Delta c_2(\mathcal{A}, \mathcal{F}).\mathcal{B} + \mathcal{G} \circ (\mathcal{A} + \mathcal{F})$ where
$\Delta c_2(\mathcal{A}, \mathcal{F}).\mathcal{B} \downarrow C_k^{\#}(E_1; E_3)$ by (11.2.13) and $\mathcal{G} \circ (\mathcal{A} + \mathcal{F}) \downarrow C_k^{\#}(E_1; E_3)$ by (11.2.10); hence c is equably continuous.

Since $\mathbb{V}.c(\mathcal{A}\times\mathcal{B}) = \mathbb{V}.\mathcal{B}\circ\mathcal{A} = (\mathbb{V}\mathcal{B})\circ\mathcal{A} \downarrow C_k^{\#}(E_1;E_3)$ by (11.2.10), c is quasi-bounded.

(B) Suppose the theorem for p and consider the map

$$c: C_k^{\#}(E_1;E_2) \times C_{k+p+2}^{\#}(E_2;E_3) \longrightarrow C_k^{\#}(E_1;E_3)$$

We assert:

(11.2.22) $$c' : C_k^{\#}(E_1;E_2) \times C_{k+p+2}^{\#}(E_2;E_3) \longrightarrow$$
$$L^{\#}\big(C_k^{\#}(E_1;E_2) \times C_{k+p+2}^{\#}(E_2;E_3);\ C_k^{\#}(E_1;E_3)\big) \text{ is } C_p.$$

Observe first that the map exists by (11.2.7), (2.6.2) and (2.9.1).

We use some abbreviations:

$$E_4 = C_k^{\#}(E_1;E_2),$$
$$E_5 = C_{k+p+2}^{\#}(E_2;E_3),$$
$$E_6 = C_k^{\#}(E_1;E_3),$$
$$E_7 = C_{k+p+1}^{\#}\big(E_2;L^{\#}(E_2;E_3)\big),$$
$$E_8 = C_k^{\#}(E_1;L^{\#}(E_2;E_3)).$$

We denote by

$\pi_1 : E_4 \times E_5 \longrightarrow E_4$ and by

$\pi_2 : E_4 \times E_5 \longrightarrow E_5$ the projections, by

$\pi_1^* : L^{\#}(E_4;E_7) \longrightarrow L^{\#}(E_4\times E_5;E_7)$ and by

$\pi_2^* : L^{\#}(E_5;E_7) \longrightarrow L^{\#}(E_4\times E_5;E_7)$ the associated linear maps.

Further by $\bar{c} : E_4 \times E_7 \longrightarrow E_8$ the composition map, and as before by $c_2 : E_4 \longrightarrow L^{\#}(E_5;E_6)$ the mapping $f \longmapsto f^*$.

Finally $D : E_5 \longrightarrow E_7$ denotes the map defined by $D(f) = f'$ and $id : E_4 \longrightarrow E_4$ the identity.

The evaluation map $e : L^{\#}(E_2;E_3) \times E_2 \longrightarrow E_3$ is bilinear and continuous by (6.2.3). Hence by (11.1.1) and (2.9.2) we have $\tilde{e} \in L^{\#}(E_8, E_4; E_7)$. By (6.4.11) we have

$$L^{\#}(E_8, E_4; E_7) \approx L^{\#}(E_8; L^{\#}(E_4;E_7)).$$

We denote by $\tilde{e}_1$ the element thus corresponding to $\tilde{e}$.

π_1, π_2 and id are obviously linear and continuous. π_1^*, π_2^* are continuous by (6.3.3), D is linear and continuous by (10.2.3).

$\bar{c}$ is C_p by our induction hypothesis. c_2 is C_p by (11.2.20).

A linear map is obviously C_p. Hence also idxD by (4.4.2).

As a consequence of the formula $c'(f,g).(\varphi,\psi) = \tilde{e}\,(g'\circ f,\varphi) + \psi\circ f$ (see (11.2.7)) we have $c' = \pi_1^* \circ \tilde{e}_1 \circ \bar{c} \circ (id \times D) + \pi_2^* \circ c_2 \circ \pi_1$ which is the composite of C_p-mappings, hence (11.2.22) by (10.4.7).

We further assert that

(11.2.23) $$c : E_4 \times E_5 \longrightarrow E_6 \text{ is } C_1.$$

From (11.2.7) we conclude that the mapping

$c : C_k(E_1;E_2) \times C_{k+p+2}(E_2;E_3) \longrightarrow C_k(E_1;E_3)$ is differentiable.

The spaces are all admissible by (10.2.4).

The mapping $c' : E_4 \times E_5 \longrightarrow L(E_4 \times E_5; E_6)$ is C_o by (11.2.22), (10.1.4) and (2.6.3). We have verified the assumptions of (11.2.8). Hence $c : E_4 \times E_5 \longrightarrow E_6$ is differentiable by (2.6.4). $c' : E_4 \times E_5 \longrightarrow L^*(E_4 \times E_5; E_6)$ being C_o by (11.2.22) and (10.1.4), we get (11.2.23).

Applying (11.4.1) to (11.2.22) and (11.2.23) one completes the induction proof of the theorem.

(11.2.24) Lemma. Let $f: E_o \longrightarrow E = \text{proj.} \lim_{i \in I} E_i$.

If $i_k \circ f : E_o \longrightarrow E_k$ is C_p for each $k \in I$, then

$$f : E_o \longrightarrow E \text{ is } C_p.$$

Proof. Let $p = 0$; $\mathfrak{W}\mathcal{A}, \mathfrak{X} \downarrow E_o$. Then $\mathfrak{W}(i_k \circ f)(\mathcal{A}) = i_k(\mathfrak{W}f(\mathcal{A})) \downarrow E_k$ for each $k \in I$. Hence $\mathfrak{W}f(\mathcal{A}) \downarrow E$. Similarly one shows that $\Delta f(\mathcal{A}, \mathfrak{X}) \downarrow E$.

Suppose the lemma for p and assume: $i_k \circ f : E_o \longrightarrow E_k$ is C_{p+1} for each $k \in I$. Then by (10.1.3) $(i_k \circ f)' = i_k \circ f' : E_o \longrightarrow L^*(E_o; E_k)$ is C_p for each $k \in I$.

Since the inclusion $i_k : E \longrightarrow E_k$ is linear and continuous, so is $(i_k)_* : L(E_o; E) \longrightarrow L(E_o; E_k)$ by (6.3.3), and also $(i_k)_* : L^*(E_o; E) \longrightarrow L^*(E_o; E_k)$ by (2.9.1). We assert

(11.2.25) $$L(E_o, E) = \text{proj.} \lim_{i \in I} L(E_o; E_i).$$

One first verifies that the underlying sets are the same. We remark however, that the projective system $L(E_o;E_i)$, $i \in I$, is slightly more general than those considered in 2.3, since the maps $j_{k\ell} : L(E_o;E_k) \longrightarrow L(E_o;E_\ell)$ induced by the inclusions

$$i_{k\ell}: E_k \longrightarrow E_\ell \quad , \quad k \geq \ell,$$

are not inclusions in the strict sense. This implies that the underlying set of $\text{proj.}\lim_{i \in I} L(E_o;E_i)$ is not the intersection of the sets $L(E_o;E_i)$, $i \in I$, but has to be constructed in the usual manner. For this one verifies that the maps $j_{k\ell}$ satisfy the transitivity condition $j_{\ell m} \circ j_{k\ell} = j_{km}$ and that they are continuous. Both conditions are easily verified, because $j_{k\ell} = (i_{k\ell})_*$. The structure of $\text{proj.}\lim_{i \in I} L(E_o;E_i)$ is the coarsest for which the induced maps $j_k = (i_k)_* : \text{proj.}\lim_{i \in I} L(E_o;E_i) \longrightarrow L(E_o;E_k)$ are continuous.

Furthermore: $\mathcal{L} \downarrow L(E_o;E) \Longleftrightarrow \mathcal{L}(\mathcal{A}) \downarrow E$ for $\forall \mathcal{A} \downarrow E_o \Longleftrightarrow$

$i_k(\mathcal{L}(\mathcal{A})) = (i_k \circ \mathcal{L})(\mathcal{A}) \downarrow E_k$ for $k \in I$ and $\forall \mathcal{A} \downarrow E_o \Longleftrightarrow$

$i_k \circ \mathcal{L} = (i_k)_*(\mathcal{L}) \downarrow L(E_o;E_k)$ for $k \in I$.

This proves that also the structures of $L(E_o;E)$ and $\text{proj.}\lim_{i \in I} L(E_o;E_i)$ are the same, hence (11.2.25). By appendix (5), (9.1.1) and (7.4.6) we get $L^*(E_o;E) = \text{proj.}\lim_{i \in I} L^*(E_o;E_i)$.

Applying the induction hypothesis to $i_k \circ f' : E_o \longrightarrow L^*(E_o;E_k)$ one concludes: $f' : E_o \longrightarrow L^*(E_o;E)$ is C_p.

It remains to show that $f\colon E_0 \longrightarrow E$ is differentiable. We have by assumption $(i_k \circ f)'(x) = i_k \circ f'(x) \in L^{\#}(E_0;E_k)$ and $R(i_k \circ f)(x) \in R(E_0;E_k)$ for each $k \in I$. Let $\mathfrak{X} \downarrow E_0$. Then $(i_k \circ f'(x)).\mathfrak{X} = i_k(f'(x).\mathfrak{X}) \downarrow E_k$ for each $k \in I$, hence $f'(x).\mathfrak{X} \downarrow E$ and thus $f'(x) \in L^{\#}(E_0;E)$.

Let $\mathcal{W}\mathcal{A} \downarrow E_0$. Then (cf. proof of (3.1.5))

$$\Theta R(i_k \circ f)(x).(\mathcal{W},\mathcal{A}) = i_k(\Theta Rf(x).(\mathcal{W},\mathcal{A})) \downarrow E_k$$

for each $k \in I$. It follows that $\Theta Rf(x).(\mathcal{W},\mathcal{A}) \downarrow E$, hence $Rf(x) \in R(E_0;E)$.

Since f' is C_0 by the induction hypothesis and (10.1.3), the assertion of the lemma is a consequence of (10.4.1).

(11.2.26) <u>Theorem.</u> The composition map

$$c\colon C_{\infty}^{\#}(E_1;E_2) \times C_{\infty}^{\#}(E_2;E_3) \longrightarrow C_{\infty}^{\#}(E_1;E_3) \text{ is } C_{\infty}.$$

<u>Proof.</u> By appendix (5), (10.2.4) and the definition given in 10.3 we have $C_{\infty}^{\#}(E_1;E_2) = \operatorname{proj.lim}_{k \in \mathbb{N}^0} C_k^{\#}(E_1;E_3)$. The inclusion $C_{\infty}^{\#}(E_1;E_2) \times C_{\infty}^{\#}(E_2;E_3) \subset C_k^{\#}(E_1;E_2) \times C_{k+p+1}^{\#}(E_2;E_3)$ is continuous and thus by (11.2.21) $c\colon C_{\infty}^{\#}(E_1;E_2) \times C_{\infty}^{\#}(E_2;E_3) \longrightarrow C_k^{\#}(E_1;E_3)$ is C_p for each $p, k \in \mathbb{N}^0$. From the lemma (11.2.21) one concludes $c\colon C_{\infty}^{\#}(E_1;E_2) \times C_{\infty}^{\#}(E_2;E_3) \longrightarrow C_{\infty}^{\#}(E_1;E_3)$ is C_p for each $p \in \mathbb{N}^0$, hence the assertion of the theorem.

§ 12. DIFFERENTIABLE DEFORMATION OF DIFFERENTIABLE MAPPINGS.

12.1 The differentiability of the evaluation map.

(12.1.1) Lemma. The mapping

$u_1 : C_o^*(\mathbb{R};E) \longrightarrow E$ defined by $u_1(f) = f(1)$ is linear and continuous.

Proof. The linearity is obvious. Furthermore $\mathbb{V}.1 \downarrow \mathbb{R}$, hence the result.

(12.1.2) Lemma. The mapping

$u: E \longrightarrow C_o^*(\mathbb{R};E)$ defined by $u(x).\lambda = \lambda.x$ is linear and continuous.

Proof. By (6.4.2) and (9.1.1) we have $E \approx L^*(\mathbb{R};E)$. Obviously $L^*(\mathbb{R};E) \subset C_o^*(\mathbb{R};E)$, and $L^*(\mathbb{R};E)$ has the structure induced by its inclusion in $C_o^*(\mathbb{R};E)$. Hence the result by (2.3.6).

(12.1.3) Theorem. The evaluation map

$e: C_{k+1}^*(E_1;E_2) \times E_1 \longrightarrow E_2$ is C_k.

Proof. The composition map

$$\hat{c}: C_{k+1}^*(E_1;E_2) \times C_o^*(\mathbb{R};E_1) \longrightarrow C_o^*(\mathbb{R};E_2) \quad (*)$$

is C_k by (11.2.21). Hence $e = u_1 \circ \hat{c} \circ (id \times \tilde{u})$ is C_k by (12.1.1), (12.1.2) and (10.4.7).

(*) $\hat{c}(f,g) = c(g,f) = f \circ g$

Because the inclusion $C_\infty^\#(E_1;E_2) \subset C_{k+1}^\#(E_1;E_2)$ is continuous for each $k \in \mathbb{N}^0$ we get obviously:

(12.1.4) <u>Theorem</u>. The evaluation map

$$e: C_\infty^\#(E_1;E_2) \times E_1 \longrightarrow E_2 \text{ is } C_\infty .$$

12.2 The linear homeomorphism $C_\infty^\#(E_1;C_\infty^\#(E_2;E_3)) \approx C_\infty^\#(E_1 \times E_2;E_3)$.

(12.2.1) <u>Lemma</u>. Let $s: E \times E \longrightarrow E$ denote the map defined by $s(x,y) = x+y$ and $\tilde{s}: E \longrightarrow C_\infty^\#(E;E)$ the map caracterized by $\tilde{s}(x).y = x+y$. Then $\tilde{s}$ is C_∞ .

<u>Proof</u>. First we show that $\tilde{s}(x)$ is an element of $C_\infty(E;E)$. But this is immediate because $\tilde{s}(x)$ is a translation for each $x \in E$. Furthermore $\tilde{s}(x+h) - \tilde{s}(x) = \bar{h}$ where $\bar{h}(y) = h$ for each $y \in E$. The mapping $h \longmapsto \bar{h}$ of E into $C_\infty^\#(E;E)$ is linear and continuous. $\bar{h}$ is a constant map for each $h \in E$, hence obviously an element of $C_\infty(E;E)$. This shows that $\tilde{s}$ is differentiable throughout E and that $\tilde{s}'(x).h = \bar{h}$ for each $x \in E$. Consequently $\tilde{s}'$ is a constant map of E into $L(E;C_\infty^\#(E;E))$ and thus C_∞ .

(12.2.2) <u>Lemma</u>. The mapping

$$c_1: C_\infty^\#(E_2;E_3) \longrightarrow C_\infty^\#(C_\infty^\#(E_1;E_2);C_\infty^\#(E_1;E_3))$$

defined by $c_1(g) = g_*$ is linear and continuous.

<u>Proof</u>. The linearity of c_1 is obvious. g_* is an element of $C_\infty(C_\infty^\#(E_1;E_2) \times C_\infty^\#(E_1;E_3)$ by (11.2.26) since g_* is a partial mapping of the composition map. Let $g \downarrow C_\infty^\#(E_2;E_3)$. We assert:

(12.2.3) $$g_* \downarrow C_k(C_\infty^\#(E_1;E_2);C_\infty^\#(E_1;E_3)) \text{ for each } k \in \mathbb{N}^0.$$

Let $k = 0$, $\forall \mathcal{A} \downarrow C_\infty^\#(E_1;E_2)$. Then $g_*(\mathcal{A}) = g \circ \mathcal{A} = \Delta c(\mathcal{A} \times 0, 0 \times g) \downarrow C_\infty^\#(E_1;E_3)$ by (11.2.25).

Suppose now the assertion for k. By (10.2.3), (10.3.3) and (2.9.1) we have $g' \downarrow C_\infty^\#(E_2;L^\#(E_2;E_3))$. From the induction hypothesis one deduces

$$(g')_* \downarrow C_k(C_\infty^\#(E_1;E_2);C_\infty^\#(E_1;L^\#(E_2;E_3))).$$

The evaluation map $e: L^\#(E_2;E_3) \times E_2 \longrightarrow E_3$ is continuous by (6.3.3). Using (11.1.1) and (2.9.2) one easily verifies that

$$\tilde{e}: C_\infty^\#(E_1;L^\#(E_2;E_3)) \times C_\infty^\#(E_1;E_2)) \longrightarrow C_\infty^\#(E_1;E_3)$$

is continuous. The same is true for the corresponding map

$$\tilde{e}_1: C_\infty^\#(E_1;L^\#(E_2;E_3)) \longrightarrow L^\#(C_\infty^\#(E_1;E_2);C_\infty^\#(E_1;E_3))$$

(cf.(6.4.11)). From (10.4.3) and (11.2.7) one concludes that

$$(g_*)' = (\tilde{e}_1)_*((g')_*) \downarrow C_k(C_\infty^\#(E_1;E_2);L^\#(C_\infty^\#(E_1;E_2);C_\infty^\#(E_1;E_3))).$$

Further we had (case $k = 0$): $g_* \downarrow C_0(C_\infty^\#(E_1;E_2);C_\infty^\#(E_1;E_3))$.

By (10.4.2) it follows that $g_* \downarrow C_{k+1}(C_\infty^\#(E_1;E_2);\ C_\infty^\#(E_1;E_3))$ which finishes the induction proof of (12.2.3). From (10.3.2) and (2.9.1) one concludes the assertion of the lemma.

(12.2.4) Lemma. If $f \in C^{\#}_{\infty}(E_1;E_2)$, then

$$f^{*} \in L\big(C^{\#}_{\infty}(E_2;E_3);C^{\#}_{\infty}(E_1;E_3)\big).$$

Proof. Let $g \downarrow C^{\#}_{\infty}(E_2;E_3)$, then $f^{*}(g) = g\circ f \downarrow C^{\#}_{\infty}(E_1;E_3)$ by (11.2.26), because $g\circ f = \Delta c(f\times 0, 0\times g)$ and c is equably continuous.

(12.2.5) Theorem. There is a natural linear homeomorphism

$$\Psi: C^{\#}_{\infty}(E_1\times E_2;E_3) \longrightarrow C^{\#}_{\infty}\big(E_1;C^{\#}_{\infty}(E_2;E_3)\big),$$ the map Ψ being caracterized by $(\Psi g)(x_1).x_2 = g(x_1,x_2)$.

Proof. Let $i_1: E_1 \longrightarrow E_1\times E_2$ resp. $i_2: E_2 \longrightarrow E_1\times E_2$ be the injections $x_1 \longmapsto (x_1,0)$ resp. $x_2 \longmapsto (0,x_2)$. Then $g(x_1,x_2) = g\big(i_1(x_1) + i_2(x_2)\big) = \big(g\circ\tilde{s}(i_1(x_1))\circ i_2\big)(x_2)$ where $s: (E_1\times E_2) \times (E_1\times E_2) \longrightarrow E_1\times E_2$ is the map discussed in the lemma (12.2.1). Hence $(\Psi g)(x_1) = g\circ\tilde{s}(i_1(x_1))\circ i_2 = (i_2^{*}\circ g_{*}\circ\tilde{s}\circ i_1)(x_1)$ where $i_1 : E_1 \longrightarrow E_1\times E_2$ is linear and continuous, $\tilde{s}: E_1\times E_2 \longrightarrow C^{\#}_{\infty}(E_1\times E_2;E_1\times E_2)$ is C_{∞} by (12.2.1), $g_{*}: C^{\#}_{\infty}(E_1\times E_2;E_1\times E_2) \longrightarrow C^{\#}_{\infty}(E_1\times E_2;E_3)$ is C_{∞} which follows easily from (11.2.21), g_{*} being a partial mapping of the composition map; finally $i_2^{*} : C^{\#}_{\infty}(E_1\times E_2;E_3) \longrightarrow C^{\#}_{\infty}(E_2;E_3)$ is C_{∞} for the same reasons. By (10.4.7) it follows that Ψg is an element of $C_{\infty}(E_1;C^{\#}_{\infty}(E_2;E_3))$.

Let

$$E_4 = C^{\#}_{\infty}(E_1\times E_2;E_3),$$

$$E_5 = C^{\#}_{\infty}(E_1\times E_2;E_1\times E_2),$$

$$E_6 = C^{\#}_{\infty}\big(E_1;C^{\#}_{\infty}(E_2;E_3)\big).$$

Then $\Psi = (i_2^*)_* \circ (\tilde{s} \circ i_1)^* \circ c_1$ where $c_1 \in L(E_4; C_\infty^*(E_5;E_4))$ by (12.2.2), $(\tilde{s} \circ i_1)^* \in L(C_\infty^*(E_5;E_4); C_\infty^*(E_1;E_4))$ by (12.2.4), $(i_2^*)_* \in L(C_\infty^*(E_1;E_4);E_6)$ by (10.4.3), (2.9.1) and the same arguments as in the proof of (11.2.26). Hence Ψ is the composite of linear and continuous maps and thus obviously linear and continuous.

Conversely let $\Phi : C_\infty^*(E_1; C_\infty^*(E_2;E_3)) \longrightarrow C_\infty^*(E_1 \times E_2; E_3)$ denote the linear map defined by $(\Phi f)(x_1,x_2) = f(x_1).x_2$. Then $\Phi f = e \circ [f \circ \pi_1, \pi_2]$ where $e: C_\infty^*(E_2;E_3) \times E_2 \longrightarrow E_3$ is the evaluation map and π_1 resp. π_2 are the projections of $E_1 \times E_2$ into E_1 resp. E_2. Hence by (10.4.5) and (10.4.7) Φf is an element of $C_\infty(E_1 \times E_2;E_3)$.

We shall prove the continuity of the mapping Φ by the formula $\Phi = e_* \circ \alpha \circ \tilde{s}(i_2(\pi_2)) \circ i_1 \circ \pi_1^*$ and using the abbreviations

$$E_6 = C_\infty^*(E_1; C_\infty^*(E_2;E_3)),$$

$$E_7 = C_\infty^*(E_1 \times E_2; C_\infty^*(E_2;E_3)),$$

$$E_8 = C_\infty^*(E_1 \times E_2;E_2),$$

$$E_9 = C_\infty^*(E_1 \times E_2; C_\infty^*(E_2;E_3) \times E_2).$$

One has $\pi_1^* \in C_\infty(E_6;E_7)$ by (12.2.4), $i_1 \in C_\infty(E_7;E_7 \times E_8)$, $\tilde{s}(i_2(\pi_2)) \in C_\infty(E_7 \times E_8;E_7 \times E_8)$ by (12.2.1), $\alpha \in C_\infty(E_7 \times E_8;E_9)$ by (10.4.5) and $e_* \in C_\infty(E_9; C_\infty^*(E_1 \times E_2;E_3))$ by (12.1.4) and (11.2.26).

From (10.4.7) it follows that Φ is C_{∞} and thus obviously continuous.

Furthermore $\Phi \circ \Psi$ and $\Psi \circ \Phi$ are the corresponding identities which ends the proof.

APPENDIX

(1) Proposition. If E is an admissible vector space, then the scalar multiplication: $\mathbb{R} \times E \longrightarrow E$ is equably continuous.

Proof. Let $\mathfrak{X} \downarrow E$. Then by (2.1.1), (2.4.2) and (7.1.1)

$$\mathcal{Y} = (\mathfrak{X} \vee (-\mathfrak{X}))^{\wedge} \downarrow E.$$

If $Y \in \mathcal{Y}$, then $Y \supset (X \cup (-X))^{\wedge}$, $X \in \mathfrak{X}$. Thus for any $x \in X$ the segment $[-x,x] \subset (X \cup (-X))^{\wedge}$ and consequently $\lambda x \in (X \cup (-X))^{\wedge}$ for each $|\lambda| \leq 1$ and $x \in X$, hence $I_1 X \subset Y$ where $I_1 = [-1,1]$, and therefore $I_1 \mathfrak{X} \leq \mathcal{Y} \downarrow E$. By (2.4.2) also $\beta . I_1 \mathfrak{X} = I_\beta \mathfrak{X} \downarrow E$ for each $\beta \neq 0$. This proves the equable continuity of the scalar multiplication by (2.8.8) and (2.5.1).

(2) Proposition. Let $\mathfrak{X}$ be a filter on a vector space. Then the filter $\mathfrak{X}^* = \mathcal{V} \cdot \sup_{\delta \neq 0} (\delta \cdot \mathfrak{X})$ is the finest one among all equable filters coarser then $\mathfrak{X}$.

Proof. Obviously $\mathfrak{X}^*$ is equable. Let $M \in \mathfrak{X}^*$. Then $M \supset V \cdot A$, where $V \in \mathcal{V}$, $A \in \sup_{\delta \neq 0} (\delta \mathfrak{X})$. Choose $\alpha \in V$, $\alpha \neq 0$. Since $A \in \delta \mathfrak{X}$ for all $\delta \neq 0$, we have $\alpha \cdot A \in \mathfrak{X}$, and therefore $M \in \mathfrak{X}$, because $M \supset \alpha A$. This proves that $\mathfrak{X}^*$ is coarser then $\mathfrak{X}$. Let now $\mathcal{Y}$ be any equable filter coarser then $\mathfrak{X}$: $\mathfrak{X} \leq \mathcal{Y} = \mathcal{V}\mathcal{Y}$

Then, for $\delta \neq 0$:

$$\delta \mathfrak{X} \leq \delta \mathcal{Y} = \delta (\mathcal{V} \mathcal{Y}) = (\delta \mathcal{V}) \mathcal{Y} = \mathcal{V} \cdot \mathcal{Y} = \mathcal{Y} \; ; \quad \text{hence}$$

$\sup_{\lambda \neq 0} (\lambda \mathfrak{X}) \leq \mathcal{Y}$ and thus $\mathfrak{X}^* = \mathcal{V} \cdot \sup_{\lambda \neq 0} (\lambda \mathfrak{X}) \leq \mathcal{V}\mathcal{Y} = \mathcal{Y}$,

which completes the proof.

Corollary. For any pseudo-topological vector space E one has:

$$\mathfrak{X} \downarrow E^* \Longleftrightarrow \mathfrak{X}^* \downarrow E.$$

(3) Proposition. Let E be any pseudo-topological vector space. Then

$$E \text{ equable} \Longleftrightarrow (\mathfrak{X} \downarrow E \text{ implies } \sup_{\lambda \neq 0} \lambda \mathfrak{X} \downarrow E).$$

Proof. (a) Suppose E equable, and let $\mathfrak{X} \downarrow E$. Then there exists $\mathcal{Y}$ with $\mathfrak{X} \leq \mathcal{Y} = \mathcal{V} \cdot \mathcal{Y} \downarrow E$. Since $\mathfrak{X}^*$ is coarser then $\mathfrak{X}$ and equable, $\lambda \cdot \mathfrak{X} \leq \lambda \cdot \mathfrak{X}^* = \mathfrak{X}^*$ for $\lambda \neq 0$; hence $\sup_{\lambda \neq 0} (\lambda \mathfrak{X}) \leq \mathfrak{X}^* \leq \mathcal{Y}$. showing that $\sup_{\lambda \neq 0} \lambda \mathfrak{X} \downarrow E$.

(b) Suppose the condition satisfied, and let $\mathfrak{X} \downarrow E$. Then we have $\mathfrak{X} \leq \mathfrak{X}^* = \mathcal{V} \cdot \mathfrak{X}^* \downarrow E$, which proves that E is equable.

(4) Proposition. Let E_i, $i \in I$ be a projective system of equable and admissible vector spaces (cf.(2.3.5)(c)).

Then $E' = \text{proj.lim}_{i \in I} E_i$ is equable and admissible.

Proof. The admissibility is proved in (7.3.2). Let $\mathfrak{X} \downarrow E$. Then $i_k(\mathfrak{X}) \downarrow E_k$ for each $k \in I$. But $\sup_{\lambda \neq 0} \lambda i_k(\mathfrak{X}) = i_k(\sup_{\lambda \neq 0} \lambda \mathfrak{X})$, since one has $\bigcup_{\lambda \neq 0} \lambda\ i_k(X_\lambda) = \bigcup_{\lambda \neq 0} i_k(\lambda X_\lambda) = i_k(\bigcup_{\lambda \neq 0} \lambda X_\lambda)$ for each such union. Hence $\sup_{\lambda \neq 0} \lambda \mathfrak{X} \downarrow E$ and thus E is equable by the preceding proposition.

(5) <u>Proposition</u>. For each projective system of admissible vector spaces, $(\underset{i \in I}{\text{proj.lim}}\, E_i)^* = \underset{i \in I}{\text{proj.lim}}\, E_i^*$.

<u>Proof</u>. (a) Let $\mathfrak{X} \downarrow (\underset{i \in I}{\text{proj.lim}}\, E_i)^*$. Then there is $\mathfrak{Y}$ with $\mathfrak{X} \leq \mathfrak{Y} = \vee \mathfrak{Y} \downarrow \underset{i \in I}{\text{proj.lim}}\, E_i$. Since $i_k(\mathfrak{X}) \leq i_k(\mathfrak{Y}) = i_k(\vee \mathfrak{Y}) = \vee i_k(\mathfrak{Y})$, one gets $i_k(\mathfrak{X}) \downarrow E_k^*$ for each $k \in I$ and thus $\mathfrak{X} \downarrow \underset{i \in I}{\text{proj.lim}}\, E_i^*$.

(b) If $\mathfrak{X}$ $\underset{i \in I}{\text{proj.lim}}\, E_i^*$, then $i_k(\mathfrak{X}) \downarrow E_k$ for each $k \in I$ by (2.6.3). But by (4) and (2.6.1) we can suppose that $\mathfrak{X}$ is equable, hence $\mathfrak{X} \downarrow (\underset{i \in I}{\text{proj.lim}}\, E_i)^*$.

NOTATIONS

$\mathfrak{A}, \mathfrak{F}, \mathfrak{X}$	filters 1.1
Φ	empty set 1.1
$[B],[A],[a]$	generated filters 1.1
$\mathfrak{X}_1 \leq \mathfrak{X}_2$	comparison of filters (1.2.1)
$\mathfrak{X} \downarrow_x E$	$\mathfrak{X}$ converges to x on E 2.1
$\mathfrak{X} \downarrow E$	$\mathfrak{X}$ converges to zero on E 2.4
$\{\mathfrak{X}_i\}_{i \in I}$	family of filters 1.2
$\sup_{i \in I} \mathfrak{X}_i$	least upper bound of filters 1.2
$\mathfrak{X}_1 \vee \mathfrak{X}_2$	sup $(\mathfrak{X}_1, \mathfrak{X}_2)$ 1.2
$\mathfrak{U}_x$	$\sup_{\mathfrak{X} \downarrow_x E} \mathfrak{X}$ (2.4.3)
$\mathfrak{X}^o$	2.7
$\hat{\mathfrak{X}}, (\mathfrak{X})^\wedge$	2.7
$\bar{\mathfrak{X}}, (\mathfrak{X})^-$	(5.3.3)
E, E_1, E_2	pseudo topological spaces
$\underline{E}$	underlying set
E^*	equable space associated to E 2.6
E^o	locally convex space associated to E 2.7
$E_1 \leq E_2$	comparison of structures 2.3
$E_1 \times E_2, \times_{i \in I} E_i$	direct product of pseudo-topological spaces 2.3

$\mathfrak{X}_1 \times \mathfrak{X}_2$	direct product of filters (on a direct product of two sets) 1.4
$\mathbb{R}$	the reals (with the natural topology)
$\mathcal{V}$	neighborhood filter of zero in $\mathbb{R}$
$[\alpha, \beta]$	closed interval in $\mathbb{R}$
I_δ	closed interval $[-\delta, \delta]$, $\delta > 0$
$\mathbb{N}$	$\{1,2,3,\ldots\}$
$\mathbb{N}^0$	$\{0,1,2,\ldots\}$
$\text{proj.lim}_{i \in I} E_i$	projective limit 2.3
$R(E_1;E_2)$	set of remainders (3.1.2)
$L_n(E_1;E_2)$	space of n-linear maps (6.1.6)
$C_k(E_1;E_2)$	space of C_k-mappings 10.2
$C_\infty(E_1;E_2)$	space of C_∞-mappings 10.3
$L_n^*(E_1;E_2)$	instead of $(L_n(E_1;E_2))^*$ (6.1.7)
$C_k^*(E_1;E_2)$	instead of $(C_k(E_1;E_2))^*$ (6.1.7)
$C_\infty^*(E_1;E_2)$	instead of $(C_\infty(E_1;E_2))^*$
$f : E_1 \longrightarrow E_2$	mapping f of E_1 into E_2
$x \longmapsto y$	x is sent into y under the considered (anonymous) map
$[f_1,f_2]$, $\prod_{i \in I} f_i$	1.3
$f_1 \times f_2$, $\times_{i \in I} f_i$	1.3

π_k	k-th projection 2.3
c	composition map
e	evaluation map
$\tilde{u}$	map associated to u (11.1.1)
$u \cdot x$	instead of $u(x)$
$\Delta f(x,y)$	abbreviation of $f(x+y)-f(x)$
$\Theta f(\lambda,x)$	abbreviation of $\begin{cases} \frac{f(\lambda x)}{\lambda} & \text{for } \lambda \neq 0 \\ 0 & \text{for } \lambda = 0 \end{cases}$
$f^{\bullet}(a)$	differential quotient (4.3.2)
$f'(a)$, $Df(a)$	derivative of f at the point a (3.2.2)
$Rf(a)$	remainder belonging to a map f which is differentiable at a (3.2.4)
$f^{(k)}(a)$	see 9.2
$D_1f(a_1,a_2), D_2f(a_1,a_2)$	partial derivatives 8.1
f^*	$f^*(g) = g \circ f$
f_*	$f_*(g) = f \circ g$
$\Longrightarrow$	implies
$\Longleftrightarrow$	if and only if
$\approx$	linearly homeomorphic

I N D E X

REFERENCES

[1] Bastiani A. : "Applications différentiables et variétés différentiables de dimension infinie", Journal d'Analyse Mathématique XIII (1964) p.1-114.

[2] Binz E. : "Ein Differenzierbarkeitsbegriff in limitierten Vektorräumen", Comm. Math. Helv. 41 (to appear).

[3] Dieudonné J. : "Foundations of modern analysis", Academic Press 1960.

[4] Fischer H.R. : "Limesräume", Math. Annalen 137 (1959) p.269-303.

[5] Keller H.H. : "Räume stetiger multilinearer Abbildungen als Limesräume", Math. Annalen 159 (1965) p.259-270.

[6] Keller H.H. : "Differenzierbarkeit in topologischen Vektorräumen", Comm. Math. Helv. 38 (1964) p.308-320.

[7] Keller H.H. : "Über Probleme die bei einer Differentialrechnung in topologischen Vektorräumen auftreten", Nevanlinna Festband, Springer (to appear).

[8] Köthe G. : "Topologische lineare Räume", Springer 1960.

[9] Kowalsky H.J.: "Topologische Räume", Birkhäuser 1961.

Offsetdruck: Julius Beltz, Weinheim/Bergstr